D0919574

INEQUALITIES FOR STOCHASTIC PROCESSES

INEQUALITIES FOR STOCHASTIC PROCESSES

(HOW TO GAMBLE IF YOU MUST)

LESTER E. DUBINS
University of California

LEONARD J. SAVAGE

DOVER PUBLICATIONS, INC.
NEW YORK

Published in Canada by General Publishing Company, Ltd., 30 Lesmill Road, Don Mills, Toronto, Ontario.
Published in the United Kingdom by Constable and Company, Ltd., 10 Orange Street, London WC 2.

This Dover edition, first published in 1976, is an unabridged, corrected republication of the work originally published by McGraw-Hill Book Company, New York, in 1965 under the title *How to Gamble If You Must: Inequalities for Stochastic Processes*. A new Bibliographic Supplement and Preface have been prepared especially for the Dover edition by Professor Dubins.

International Standard Book Number: 0-486-63283-0
Library of Congress Catalog Card Number: 75-25001

Manufactured in the United States of America
Dover Publications, Inc.
180 Varick Street
New York, N.Y. 10014

TO OUR FATHERS

IT IS ALMOST ALWAYS GAMBLING THAT ENABLES ONE TO FORM A FAIRLY CLEAR IDEA OF A MANIFESTATION OF CHANCE; IT IS GAMBLING THAT GAVE BIRTH TO THE CALCULUS OF PROBABILITY; IT IS TO GAMBLING THAT THIS CALCULUS OWES ITS FIRST FALTERING UTTERANCES AND ITS MOST RECENT DEVELOPMENTS; IT IS GAMBLING THAT ENABLES US TO CONCEIVE OF THIS CALCULUS IN THE MOST GENERAL WAY; IT IS, THEREFORE, GAMBLING THAT ONE MUST STRIVE TO UNDERSTAND, BUT ONE SHOULD UNDERSTAND IT IN A PHILOSOPHIC SENSE, FREE FROM ALL VULGAR IDEAS.

LOUIS BACHELIER

PREFACE TO THE DOVER EDITION

The occasion of this edition provides an opportunity to honor the memory of Leonard J. Savage, who suddenly died of a heart attack on November 1, 1971.

Much will certainly be written about Jimmie Savage as a man and about Jimmie Savage as a scientist. In this preface, it seems appropriate to confine myself to a few of my many recollections of our collaboration in the writing of this book.

Jimmie had already made a wonderful impression on me when I was a student at the University of Chicago in the early 1950's. He was different from the other mathematicians I had heretofore known. He often tested his understanding of the most abstract mathematics with colorful, concrete examples that seemed to leap instantly and effortlessly into his mind. He was often searching for answers, and he shared his thought processes with you. In this way, you were his equal, for the search was a joint venture—you searched too. But one could not fail to be impressed with his quickness of mind as well as with the breadth and depth of his knowledge.

When in June of 1956 I returned to Chicago after a year's absence, Jimmie amazed me utterly by telling me that he had been unable to settle the problem of how to play Red-and-Black wisely. How could so simple-sounding a problem escape his (and others') immediate solution, I wondered. So I began to think about this enticing problem. Little did I then anticipate the delightful conversations and correspondence that were to develop from that beginning, nor did I then realize the influence that it was destined to have on my own career. But to return to June 1956—

Within a few days of Jimmie's telling me of the problem of Red-and-Black, I happened to make this trivial observation: if bold play is optimal, it is not uniquely so. If I was amazed at learning that the problem of Red-and-Black was open, I was even more so at Jimmie's reaction to my all but immediate observation. He was so incredulous that he checked, and double-checked—and even at the blackboard! He made me feel good about my tiny observation by saying that he had always taken the contrary for granted. It was his manifest appreciation of that infinitesimal advance that gave me the courage to accept his invitation that I work with him. His ability to recognize and to appreciate genuinely even the tiniest of contributions greatly enhanced my pleasure in working with him during the succeeding months and years.

By the end of the summer of 1956, Red-and-Black was no longer an enigma to us. We had shown that bold play was optimal (Section 5.11). What launched this book, however, was our mistaken confidence that we could, without difficulty, extend our understanding to the more general problem of primitive casinos (Chapter 6). A plausible conjecture was that bold play was again optimal (Section 6.1). So I (and perhaps Jimmie, too) was shocked when, after we had spent more than two years in seeking a proof, Aryeh Dvoretsky discovered that bold play was not optimal (Section 6.6), at least if playing time is limited. Our—or at least *my*—confidence in the conjecture shaken, we nevertheless persevered and, within a few additional months, settled a combinatorial inequality that had stumped us for more than two years, and which was all that had remained to complete the proof that bold play was indeed optimal for subfair, primitive casinos. This was the now innocent-looking inequality of Section 6.3 which asserts that certain sequences M are Cartesian.

Another concrete problem that we found easier to pose than to settle was to determine those casinos that are, in a certain sense, fair (Section 10.3). We made some progress, were pleased by certain mildly surprising observations, and then, were much surprised by a discovery of Donald Ornstein (Section 10.2). The main problem of giving necessary and sufficient conditions for a casino to be fair was finally solved, mainly by the prodigious efforts of Jimmie during the year 1958–59, which he spent in Rome.

At what point did we break with tradition and develop the theory without the countably additive hypothesis? It happened essentially thus. In our work on special problems such as Red-and-Black, we had, like many others before us (and I mention in particular Snell, 1952), used the idea of conservation of fairness, which in this book finds important expression in the basic Theorem 2.12.1. But it was not until I had heard Aryeh Dvoretsky speak about the problem of bold play for primitive casinos (with limited playing time) that I began to appreciate the power and generality of that idea. I immediately saw the possibilities of a general theorem, and established one, but one which applied only to countably additive problems, and not even to all such. What was most mystifying and puzzling, however, was that it should require, as it did, a preliminary technical result of moderate difficulty, namely, a measurable selection theorem. It was at this point that Jimmie had the daring idea, influenced no doubt by his association with Bruno de Finetti, that we aim for a general, finitely additive theory. Once Jimmie, with some assistance from de Finetti, had persuaded me that we were not embarking on totally unjustifiable heresy, we proceeded to develop the appropriate mathematics. The main difficulty which we encountered was to determine the proper class of functions which were to be integrated by strategies, which class turned out to be the inductively integrable functions (Theorem 2.8.1). Once that was accomplished, the fundamental ingredients to a theory of gambling were quickly developed, in particular, the basic Theorem 2.12.1 and its companion, Theorem 2.14.1. The finitely additive approach made it possible to understand the true, and far less central, role of the measurable selection theorem, which theorem now appears later in Section 2.16.

An abstract problem that we raised but did not settle was the extent to which stationary families of strategies are adequate. For us, it was already a challenge to settle this for various special cases, including that in which the fortune space, F, is finite (Section 3.9). Many further developments of this problem of stationarity have taken place, as a number of entries in the Bibliographic Supplement indicate, the most surprising of which is due to Ornstein (1969).

It is almost no exaggeration to say that working with Jimmie, whether face to face or by correspondence, was always a joy for me. His letters were a delight. I chuckled over his humor and enjoyed his insights. Our work was accomplished mainly through correspondence punctuated by periods together. The letters grew into manuscripts which invariably went through several editions. The final drafts that Jimmie would write were, for me, a thing of beauty. At least once, I undertook to write the final draft of something we had done. And I took great pains, much thought and much time, until I was totally satisfied that it could not be improved upon, not even by Jimmie. But after Jimmie read my draft and then wrote his (I do not say "rewrote," for there was barely any resemblance), my chagrin gradually changed to delight and admiration at his achievements, both mathematical and expository.

More than once, Jimmie told me that he considered our work to be an important contribution, and that it would be so judged by our peers. Perhaps that is why the constant flow of calls upon his time and talents did not prevent him from pouring into our joint work so much of his energy and genius. Working thus with Jimmie for more than three years, it would have been all but impossible for anyone not to produce something of interest.

So, upon coming to Berkeley in 1959—three years and three months after Jimmie had introduced me to the problem of Red-and-Black—I was eager to communicate our results and methods, especially those mentioned above. It was a pleasure to lecture about our work, and to have an interested audience, especially in the person of David Blackwell, who suggested that we publish our findings as a book. At every stage in the writing of this book, Jimmie was an artist interested in the quality of every aspect of the book, whether it be the elegance of an argument or the appearance of a page. Though his example was inimitable, it could not fail to make his coauthor a better collaborator. Our efforts were fully rewarded with the sense of completion which came with the circulation of the mimeographed edition, dated August 28, 1960.

Each of the subsequent editions, whether mimeographed or printed, and this one, in particular, differs but little from the preceding. In order to alleviate the present publisher's concern about possible misunderstandings as to the nature of the book, its title and subtitle have been permuted for this edition. The score or so of minor mechanical errors that miraculously appeared in the last edition—and which were brought to light mainly by the participants in a seminar led by Richard Olshen and Jimmie Savage at Yale in 1967 have been corrected, of course. More important, this edition contains a Bibliographic Supplement. The inclusion of this Supplement, unannotated as it is, can merely call attention to recent advances. But reading some of the references will, I hope, enable the reader to visualize, at least in outline form, the kind of revision that might be possible were there alive today my much missed friend, Leonard Jimmie Savage.

L. E. Dubins

University of California
Berkeley, California
August 1975

PREFACE TO THE FIRST EDITION

The immediate object of this book, a chapter in the pure mathematics of probability, is to explore and to share with other mathematicians certain problems that have excited our curiosity and that lead to somewhat novel methods and phenomena. A light introduction to these problems is in Chapter 1.

In spite of the probabilist's tendency to invoke gambling imagery, works on probability have little influence, for good or bad, on the gambling man. But probability is not behind other parts of mathematics in its influence on practical affairs, and this book about theoretical problems of how to make the best of a bad and risky situation also may eventually lead to applications.

We have tried to write the book for mathematicians generally. Technical knowledge of probability is not a logical prerequisite, except for isolated passages, but it will ordinarily be a psychological one. No one may be disposed to read the whole book through, so we have tried to help skippers and postponers by providing them with signposts.

We began to ask each other how to play red-and-black in the summer of 1956, and this report of our research—in our minds first a note, then a paper, and at last a book—has been nearly finished ever since. Six written preliminary announcements are (Savage 1957), (Dubins and Savage 1960, 1960*a*), (Savage 1962), and (Dubins and Savage 1963, 1965). We also gave talks about the research, particularly at the following meetings: the Institute of Mathematical Statistics at Atlantic City in September, 1957, and at Albuquerque in

April, 1962; the American Mathematical Society at Chicago in January, 1960.

The flexible system of making references to the Bibliography by such expressions as "a book by Doob (1953)", "(Doob 1953)", or "(Doob 1953, p. 67)" is familiar and almost self-explanatory. Other mechanical matters should also be unobtrusive, but some points may be mentioned.

Technical terms are defined with the aid of italics, which usually provide the only signal that a sentence is a definition. The number of technical terms and notations introduced is large in spite of economy drives, but only a few are in use in any one section and many are short-lived. The Subject Index and Index of Symbols will lead promptly to a definition that has been forgotten or skipped.

To illustrate the not unusual system of internal references, "Theorem 2.12.1" denotes the first theorem of Section 2.12, that is, of Section 12 of Chapter 2. Within Section 2.12, it is called simply "Theorem 1", and elsewhere within Chapter 2, it is called "Theorem 12.1". Similarly, "(2.12.1)" is the full name of the first displayed formula in Section 2.12; the shorter names "(1)" and "(12.1)" are used within Section 12.1 and elsewhere in Chapter 2, respectively. This system is facilitated by showing a section number at the top of almost every pair of pages.

The mark "◆" signifies the end of a proof.

LESTER E. DUBINS

LEONARD J. SAVAGE

ACKNOWLEDGMENTS

We have benefited from conversations with many friends, whose specific contributions will be mentioned in context. The help of David Blackwell, Gerard Debreu, Bruno de Finetti, Aryeh Dvoretzky, Paul R. Halmos, Lawrence Jackson, Jesse Marcum, John Myhill, Donald Ornstein, Roger A. Purves, and W. Forrest Stinespring was particularly valuable and encouraging.

It was David Blackwell who suggested that we put our work on gambling into the form of a book. His counsel and suggestions have contributed much, and his continued interest in the project has kept it fresh and rewarding for us.

Energetic criticism by the students in a seminar based on a preliminary version has made this book better. For this, we particularly thank Vida Greenberg, Martin Helling, Lawrence Jackson, Michel Jean, and Roger A. Purves. Criticism of a late version by Bruno de Finetti, Aryeh Dvoretzky, and Thomas Ferguson initiated many improvements.

Countless sentences are clearer and more graceful, thanks to Louise Forsyth's suggestions.

This book entailed an enormous amount of loyal and skilled secretarial work, and we are grateful to all who participated in the many drafts. The final and most demanding one was prepared by Mrs. Jaynel E. Moore.

Dubins has worked on this book at the University of Chicago, the Carnegie Institute of Technology, the Institute for Advanced

Study, the University of Michigan, and the University of California at Berkeley; Savage has worked on it at the University of Chicago, the University of Michigan, Yale University, the University of California at Berkeley, and at the RAND Corporation.

Our work was given generous financial support by several agencies: the Air Force Office of Scientific Research and the National Science Foundation (through a grant and through its Regular Postdoctoral Fellowship program) for Dubins' participation; the Office of Naval Research, the Air Force Office of Scientific Research, the National Science Foundation, the John Simon Guggenheim Memorial Foundation, and the Michigan Institute of Science and Technology for Savage's. A grant from the Ford Foundation to Savage has defrayed travel and certain other expenses, thus making it more practical for us to work together.

The passage on the quotation page is from (Bachelier 1914, page 6), with the permission of the publisher, Flammarion et Cie.

CONTENTS

Each chapter leads most naturally to the next. Some chapters, however, lead fairly naturally to later ones, as this diagram indicates.

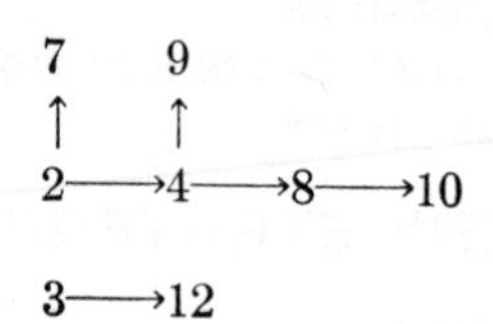

1 INTRODUCTION

1. THE PROBLEM. Imagine yourself at a casino with $1,000. For some reason, you desperately need $10,000 by morning; anything less is worth nothing for your purpose. What ought you do? The only thing possible is to gamble away your last cent, if need be, in an attempt to reach the target sum of $10,000. There may be a moment of moral confusion and discouragement. For who has not been taught how wrong and futile it is to gamble, especially when short of funds? Yet, gamble you must or forgo all chance of the great purpose that can be achieved only at the price of $10,000 payable at dawn. The question is how to play, not whether.

As is well known, any policy of compounding bets that are subfair to you must decrease your expected wealth. Consequently, no matter how you play, your chance of converting $1,000 into $10,000 will be less than 1/10. How close to 1/10 can you make it and by what strategy? That is the sort of problem this book attacks. The particular problems treated are highly simplified and idealized, especially when we get down to cases, and we do not leave our armchairs far enough to discover whether many of them would be applicable in such complicated situations as obtain for any real gambler.

The fantasy with which we have introduced the general problem of optimal gambling systems has no immediate practical importance. But the problem, once proposed, cries out for attention as pure mathematics. As such, we have found it stimulating and hope you will.

2. PREVIEW

A rough idea of some of the more colorful problems and conclusions of this book can be imparted quickly.

Suppose the only kind of gamble available to you in your effort to convert a sum of money into a larger target sum is to bet, at fixed, subfair odds, on independent repetitions of some fixed kind of event, such as drawing a red card or a spade. You are free to choose for yourself the amount to stake on each bet, subject only to the restriction that you can never stake what you do not possess. Under these circumstances, one—not usually the only—optimal strategy for you is to play *boldly;* that is, always to stake on each bet either all the money in your possession or just enough to arrive immediately at the target sum in case you win the bet, whichever of these two stakes is the smaller (Sections 5.3 and 6.3).

If your object is, as before, to reach a target sum and if you are free at each play to stake any amount within your means, dividing the stake equally among as many numbers on a roulette table as you wish, then it is ill-advised to bet on more than one number at a time (Theorem 6.9.2). This may be true even if you are free to stake different amounts on different numbers simultaneously, but that is an open and promising question.

In games that have several possible outcomes, the gambler does not merely lose his stake or win back a predetermined multiple of it, as when he simply bets on an event. Instead, he wins back some random multiple (quite possibly 0) of the stake that he has deposited, so his bet is described as a random variable—his winnings, possibly negative—with an essential minimum not less than the negative of the stake. A popular index c of unfairness of a bet is the ratio of its expected loss to its stake. What can be said about the probability of converting a fortune f into the larger target sum 1 if the house permits only bets of index at least c? The probability is at most $1 - (1 - f)^{1-c}$, which can almost be attained by, for example, repeatedly betting very small stakes on very improbable events at such odds that winning one of the bets will bring the gambler at once to the target sum 1 (Section 9.3).

A second index α of unfairness is the ratio of the expected value of the loss to the variance. The gambler's probability of converting f

into 1 with only bets of index at least α is at most $f/(1 + \alpha - \alpha f)$. This bound also is nearly attained by successively placing small stakes on the chance of leaping to the goal (Section 9.4).

A third index of unfairness β is the ratio of the expected loss to the first absolute moment. This index was introduced by Donald Ornstein, who showed that the least upper bound is $(1 - \beta)f/(1 + \beta - 2\beta f)$ and that it is attained by staking the whole of f on a single chance of attaining the goal (Section 9.2).

When a gambling situation permits you to make fair bets, you may be able to reach 1 from f with probability f. In particular, if there is available a fair game that you can repeatedly play, choosing the stake each time at your own discretion (within your means), you "ought" to be able to reach 1 with probability arbitrarily close to f. Surprisingly, though, for each positive ϵ there are fair games that will not permit you to reach 1 from f with probability greater than ϵ, as was discovered by Donald Ornstein (Section 11.2).

This book does not consist entirely of the study of such concrete and colorful problems, primarily because to formulate and solve such problems satisfactorily requires considerable abstraction and generalization. We came gradually to adopt a certain concept of finitely additive, time-discrete stochastic processes as the model of a gambling strategy. Since finitely additive measures have not been popular, the mathematical fundamentals of such processes had to be explored, and this produced some analytic novelties (Chapter 2). The abstract formulation, once achieved, suggested abstract questions (Chapters 2, 3, and 7), and it had to be related to other abstract structures already familiar to probability theory (Chapter 12).

Many concrete problems—in particular, those reported on in this section—refer to a moderately general kind of gambling house, here called a casino, associated with nonnegative stochastic processes, and these problems suggested a theory of casinos of interest in itself (Chapter 4).

Many facts about gambling can be translated with almost no loss into the language of stochastic processes. For example, the limitation on a gambler constrained to use bets for which the expected loss is at least α times the variance says scarcely more than this: If X_0, $X_1, \cdots$ is an expectation-decreasing semimartingale for which $X_n \geq 0$ for all n, $X_0 < 1$, and the conditional expectation of $X_n - X_{n+1}$

given the past is at least α times its conditional variance given the past, then $X_0/(1 + \alpha - \alpha X_0)$ is an upper bound for the probability that $\sup_n X_n \geq 1$. And this bound is sharp.

In short, one thing has led to another, as happens in mathematics, and it is not possible to say satisfactorily in advance what all the parts of the book are about. But the introduction of each chapter offers additional summary information.

3. ANTECEDENTS

Probabilists have been interested in gambling since mathematical probability began; so it is surprising that problems like the first one mentioned in the preceding section had not been solved, or even much discussed. Perhaps moral disapproval has focused so much light on those respects in which gambling is futile as to leave other aspects of gambling in darkness.

The idea that large bets are sometimes advantageous for attaining a fixed goal is in the gambling air. The earliest mathematical reference to it that we know is by Coolidge (1908–1909). He studies the first problem of the preceding section, generalized to include the possibility that there is a legal upper limit to your stake, and purports to prove that your only optimal strategy is always to stake the legal limit, or all that you possess, or enough to reach the target, whichever is least. The assertion of uniqueness is incorrect. Whether that of optimality is correct when there is a legal limit, we do not know. We have not seen how to rescue much from Coolidge's faulty demonstration.

The only other reference we know in this immediate area is a passage in which Feller (1950, p. 285, first edition only) comes to the not quite accurate conclusion that bold play is uniquely optimal when there is no legal limit.

Knowledge that small bets are contraindicated in certain situations is widespread. As De Moivre (1711, Problem IX) and James Bernoulli (1713) were quite capable of showing, if you engage in only the smallest allowable bets on red-and-black, say $1 or less, you are all but certain to lose your entire $1,000 before reaching $10,000. An interesting paper by Blackwell (1954) and a sequel to it by Weingarten (1956) can be interpreted as greatly generalizing the method discovered by De Moivre and as treating certain problems about how to gamble if you must relatively completely.

An extension of the recommendation of large bets is the recommendation of bets for which the variance is large compared with the expected loss. This idea goes back at least to de Finetti (1939, 1940), whose exposition of it we paraphrase in Section 8.7. Incidentally, our work on gambling might never have begun had one of us (Savage) not been intrigued by hearing of this rule indirectly through Jesse Marcum. De Finetti's papers (1939, 1940) are also descendants of De Moivre's Problem IX.

One very important idea about gambling strategies permeates the whole history of probability. It might be called the "conservation of fairness", and it says that any combination or concatenation of bets that are at most fair, when viewed as a single, composite bet, must be at most fair. For example, De Moivre takes the conservation of fairness for granted in his Problem IX, and we find it our most important tool.

Bachelier (1900, 1912, 1914, 1938) deliberately makes the conservation of fairness central to his theory of speculation in stocks and commodities. This work is apparently the oldest sustained study of continuous-time stochastic processes. As we have heard it put, Brownian motion, a gem of modern physics, is a botanical discovery the mathematical theory of which originated in the social sciences.

Some formulations of conservation of fairness in the measure-theoretic framework date from (Halmos 1939). These are carried forward in Doob's (1953) treatise on stochastic processes.

While we were working on this book, which is largely motivated by unfavorable games, Leo Breiman was independently working on how to get rich quick at favorable games. His interesting paper (Breiman 1961) and this book complement each other.

When the theory of gambling systems is viewed very abstractly, it tends to merge, in principle, with the theory of sequential decisions and with the general idea of dynamic programming, a broad concept of strategy optimization developed by Bellman (1952, 1957). The relation of the theory of gambling strategies to dynamic programming and to the establishing of inequalities for stochastic processes is discussed in Chapter 12.

2 FORMULATION OF THE ABSTRACT GAMBLER'S PROBLEM

1. INTRODUCTION.

This chapter defines gambler's problems abstractly and demonstrates some properties of theirs that go a long way toward solving the relatively specific cases in later chapters. The abstract formulation is not very hard to comprehend, and we believe that it helps bring out the essence of many arguments, that it is of some interest in itself, and that it may stimulate future work. Some general questions that logically pertain to the abstract formulation are postponed to later chapters (3, 7, and 12) so that you may more quickly meet the concrete cases that originally stimulated our own interest and enliven the general theory.

2. FORTUNES AND HISTORIES

A gambler begins with a *fortune* f_0, and as play progresses he moves successively through a sequence of fortunes $f_1, f_2, \cdots$.

A fortune might well be simply a sum of money in the possession of the gambler, and in the examples in most later chapters (4, 5, 6, 8, 9, 10, and 11) it can be so interpreted. There are, however, much more general possibilities of interest, which will be explained in Chapter 12; for "fortune" is so conceived that the entire worth of a situation, which means the range of choice (that is, the set of gambles) open to the gambler in making his next gamble, as well as the value to the gambler of what he will possess if he now ceases gambling, is a function of his present fortune alone. The generality envisaged is so great that the word "state" might be more appropriate for this chapter, but "fortune" is more suggestive for the concrete problems of later chapters. The purpose of this chapter and the next is to present that part of a mathematical theory of gambling which is general enough to embrace any space of fortunes. In Chapters 7 and 12 there is some additional general theory.

Let F, an arbitrary nonempty set, be the set of all fortunes and H the set of all sequences of fortunes, that is, the set of all *histories*, $h = (f_1, f_2, \cdots)$. In our conventions, f_0 does not belong to the history; it is "prehistoric".

3. GAMBLES

A *gamble* is, of course, a probability measure γ on subsets of fortunes. In the tradition of recent decades, such a measure would be defined only on a sigma-field of subsets of F and required to be countably additive on that sigma-field.

If this tradition were followed in this book, tedious technical measurability difficulties would beset the theory from the outset. (To see this kind of potential difficulty, formulate mathematically the problem corresponding to that of a gambler with an initial fortune f who desires to maximize the probability that his fortune at the end of two gambles will be in a certain subset of the unit interval, where for each g in the interval there is a set $\Gamma(g)$ of countably additive gambles defined only on the Borel subsets of the interval.) Experience and reflection have led us to depart from tradition and to assume that each γ is defined for all subsets of F and not to assume that γ is necessarily countably additive. This departure is further explained and justified in the next paragraphs.

The assumption that γ is not defined for all sets would, all in all, complicate this chapter; the restriction to countably additive gambles would weaken its conclusions. Some of the new problems that the finitely additive approach does introduce are of mathematical interest in themselves.

When a gamble is specified in practice—even in the most mathematical practice—the specification will often define the value of the gamble only on some subclass of the class of all sets, perhaps on a Boolean algebra. For example, it might be specified that a certain gamble coincides with Lebesgue measure for the Lebesgue-measurable subsets of the unit interval. It is therefore essential to handle problems in which the gambles are not defined for all subsets of fortunes. One way to do this, suggested by tradition, is to carry a concept of measurability and integrability throughout the discussion, exploring the integrability of various functions that arise as candidates for integration, and to discuss upper and lower (or outer and inner) integrals when nonintegrable functions do arise.

A seemingly equivalent and, we find, much simpler method of handling problems where gambles are defined only on a subclass of sets is to consider all extensions of each such incompletely defined gamble to the class of all sets of fortunes. According to the Hahn-Banach theorem, such extensions exist in abundance, though in a very nonconstructive sense. If, for example, the gambler starting from $1,000 can reach $10,000 with probability .07 in every completion of an originally incompletely defined problem, is it not a sensible interpretation to credit him with at least that much in connection with the problem as originally specified? Likewise, if there is something he cannot achieve (or approach) under any extension, it ought not be regarded as achievable in the original problem. Finally, if something can be approached for some extensions but not for others, then the original problem must be recognized as not sufficiently specified to yield a definite answer.

The traditional extent of specification is, as might be expected, often found to be adequate. For instance, in many interesting cases of a gambler intent upon maximizing the probability of converting a given initial cash fortune into a specified larger one, it is irrelevant how the gambles are defined on non-Borel sets, or even non-Jordan sets (Corollaries 4.6.1 and 8.2.5). Even if the space of fortunes is not an interval of real numbers, a general theorem in Section 2.16 paves the way for a theory of gambling consonant with the traditional theory of probability, but less general than the one set forth in some detail in this chapter.

With these explanations in mind, we make everywhere-defined gambles the main object of our formal study and use them as the convenient tool for studying the less completely defined gambles that are the more practical objects of interest.

De Finetti (1930, 1937, 1949, 1950, 1955, 1955*a*) has always insisted that countable additivity is not an integral part of the probability concept but is rather in the nature of a regularity hypothesis; his papers (1949) and (1950) are more particularly concerned with the conceptual aspects of this question than are the others cited; (1955) and (1955*a*) are mathematical papers about finite additivity. Personal contact with de Finetti gave us the courage to break with the traditional restrictions of countable additivity and to view countably additive measures much as one views analytic functions—as a particularly important special case.

We believe that anyone familiar with countably additive measure will need no special preparation to understand the aspects of finitely additive measure that arise in this book. One who does wish to see an explicit development of finite additivity in English can turn to (Dunford and Schwartz 1958, Sections 1 through 3 of Chapter III, and notes and remarks, p. 233).

One of the many ideas that is as applicable in the finitely additive theory as in the more special countably additive one is that of "completing" a partially defined measure or integral γ. Of course if ϕ_1 and ϕ_2 are integrable under γ with $\phi_1 \leq \phi_2$, then $\int \phi_1 \, d\gamma \leq \int \phi_2 \, d\gamma$; that is, γ is *order-preserving.* There are usually many ϕ to which γ can be extended so as to remain order-preserving. To "complete" γ is to extend it to those ϕ to which it has only one possible extension, namely, those whose upper and lower integrals are equal. Since the idea of completing a measure space does not appear to be different from "the method of exhaustion" often attributed to Eudoxus, we call the result the *Eudoxus extension* to distinguish it from further extensions. The linearity and the order-preserving character of the integral determine the value of $\int \phi \, d\gamma$ for a large class. In the event that γ is defined on all subsets of F, this class includes all bounded ϕ.

Besides those of primary interest for this book, there are other reasons for pursuing the study of finitely additive measures. To mention but one, sometimes the only natural measure is not countably additive. A natural and intuitive example that does not yet seem to be in the literature is this. There is one and only one translation-invariant probability measure defined on the Boolean algebra generated by the arithmetic sequences. Under this measure, the set $\{\cdots, -2n, -n, 0, n, 2n, \cdots\}$ is necessarily assigned a probability of $1/n$. An obvious relation between this measure and the more familiar notion of the (long-run) density of a subset of the integers is this: The upper and lower densities of a subset are between the upper and lower measures. Nathan Fine has told us interesting number-theoretic facts, not yet published, that have flowed from his study of the completion of this measure. Another finitely additive measure, suggested by de Finetti, is one that assigns to every interval of rational numbers the distance between its endpoints. Any probability measure that assigns probability 1 to some countable set, but probability 0 to every finite set, will be called *diffuse*.

4. GAMBLING HOUSE

The *gambling house* is, in effect, a function Γ that associates with each f a nonempty set $\Gamma(f)$ of gambles γ among which the gambler is allowed to choose when his fortune is f. The function Γ thus describes the environment in which the gambler plays. Often, the one-point gamble $\delta(f)$ is in $\Gamma(f)$ for all f.

5. STRATEGIES

The gambler, constrained by his initial fortune f_0 and subject to the rules imposed by the gambling house Γ, must decide how to play; he must choose one among all available strategies. Just what is a strategy? How wide is the domain of the gambler's choice? Loosely, a strategy ought to be a rule—not necessarily a good one—specifying what gamble the gambler is to choose in every possible contingency.

Formally, a *partial history* p is a finite (possibly vacuous) sequence of fortunes $(f_1, \cdots, f_n)$, and a *strategy* is a function σ associating a gamble $\sigma(p)$ with each partial history p. Equivalently, a strategy σ is a sequence $\sigma_0, \sigma_1, \cdots$, where σ_0 is a gamble, and for each positive n, σ_n is a function that associates with each finite sequence of fortunes $(f_1, \cdots, f_n)$ a gamble $\sigma_n(f_1, \cdots, f_n)$. For each subset A of F, $\sigma_n(f_1, \cdots, f_n)A$ is, conceptually, the conditional probability under σ that $f_{n+1} \in A$ given $f_1, \cdots, f_n$.

The gambler with the strategy σ who has experienced the partial history p continues gambling according to the *conditional strategy* $\sigma[p]$, where $\sigma[p](p')$ is the gamble $\sigma(pp')$ associated by σ with the partial history pp', namely the finite sequence p followed by the finite sequence p'.

The strategy conditional on a one-fortune partial history $p = (f)$ is naturally abbreviated from $\sigma[(f)]$ to $\sigma[f]$. Thus each σ determines a gamble σ_0 and a family of strategies $\sigma[f]$; and each gamble γ and *family of strategies*, or strategy-valued function of f, $\bar{\sigma}$ determine the strategy of which $\sigma_0 = \gamma$ and $\sigma[f] = \bar{\sigma}(f)$.

The house Γ and the gambler's initial fortune f limit his range of choice of strategy. Formally, σ is *available in* Γ *at* f if $\sigma_0 \in \Gamma(f)$ and $\sigma_n(f_1, \cdots, f_n) \in \Gamma(f_n)$ for $n \geq 1$.

Obvious theorems like the following will often be stated without proof or comment.

THEOREM 1. If σ is available in Γ at f', then, for every f, the strategy $\sigma[f]$ is available in Γ at f.

The technical concept of strategy adopted here is incomplete as a formulation of a full policy of behavior on the part of the gambler. It leaves out of account the idea that the gambler must, sooner or later, decide to stop playing. One possible program for introducing this feature is to confine attention to strategies for which σ_n is almost sure to be a fixed, one-point measure for sufficiently large n. We have found it more convenient to study all strategies and to combine a strategy with a stop rule to form a policy (in Section 9). One advantage of defining a policy to consist of a strategy together with a stop rule, rather than restricting attention to strategies that stagnate almost surely, is that in some problems the simplest strategies (and sometimes the only ones) that have a claim to be called optimal are not almost sure to stagnate but have to be arbitrarily terminated by a stop rule; there is a natural example in Section 11.6.

6. THE INTEGRAL GENERATED BY A STRATEGY

With respect to a strategy σ, it ought to be meaningful, according to the intended formulation, to ask, for example, for the probability that at least 3 of the first 17 fortunes fall in some subset of the set of all fortunes or (when the fortunes are positive integers) for the probability that the f_1st coordinate of the history h exceeds f_1, that is, for the probability that $f_{f_1} > f_1$. In particular, σ is intended to induce a certain probability measure, or integral, at least on bounded functions g of h that depend on only a finite number of coordinates of

h, namely,

$$(1)\quad \sigma g = \iint \cdots \int g(f_1, f_2, \cdots, f_n)\, d\sigma_{n-1}(f_n | f_1, \cdots, f_{n-1}) \times \cdots d\sigma_1(f_2 | f_1)\, d\sigma_0(f_1).$$

In ordinary parlance, σg is the expected value of g under σ. Note that the symbol "σ" is now being used both for a strategy and for the integral induced by it. Two nominally different strategies σ and σ' can induce the same integral if, for example, $\sigma_0 = \sigma_0'$ and $\sigma[f] = \sigma'[f]$ for almost all f with respect to σ_0.

The definition (1) does not go far enough to meet our intentions or our needs.

Example 1. Let F be the positive integers, and let $g(h)$ be 1 or 0 according as the f_1st coordinate of h, namely f_{f_1}, is odd or even. Since $g(h)$ depends on only the first and f_1st coordinates of h, $g(h)$ may be written as $g(f_1, f_{f_1})$. In evident notation, σg ought to be

$$(2)\qquad \sigma g = \int \sigma[f_1] g(f_1, f_{f_1})\, d\sigma_0(f_1),$$

where the meaning of the integrand for each f_1 is given by (1). But (1) alone cannot deliver this conclusion, since g is not determined by any finite number of coordinates of h. Still more, it may not be possible to approximate g usefully from above and below by functions that are so determined. For suppose that g_* and g^* depend on only the first m coordinates of h and that $g_* \leq g \leq g^*$. Plainly, $g_*(h) \leq 0$ and $g^*(h) \geq 1$ if $f_1 > m$. Therefore, according to (1), if σ_0 is a diffuse gamble, then $\sigma g_* \leq 0$ and $\sigma g^* \geq 1$. But σg as defined by (2) can, depending on the family $\sigma[f_1]$, have any value in $[0, 1]$. Thus (1) cannot deal with this g even through the technique of Eudoxus completion. For the lower integral $\sigma_* g$ is at most 0, and the upper integral $\sigma^* g$ is at least 1, so g is not Eudoxus integrable.

For each fortune f and history $h = (f_1, f_2, \cdots)$, let fh denote the history $(f, f_1, f_2, \cdots)$. Each g defined for all h gives rise to a family of functions gf, one for each f, defined for each history h thus: $gf(h) = g(fh) = g(f, f_1, f_2, \cdots)$. In the spirit of (2), it would be tempting to evaluate σg by

$$\sigma g = \int \sigma[f_1] g f_1 \, d\sigma_0(f_1), \tag{3}$$

whenever gf_1 is, for each f_1, the kind of function of h whose expectation under $\sigma[f_1]$ has already been defined. Thus (3) is promising if, for each f_1, gf_1 depends on only a finite number of coordinates, etc. The "etc." is a little delicate; a section must be devoted to its exploration before the hope expressed in (3) can be fulfilled.

7. FINITARY MAPPINGS

Much of the content of this section is already to be found in (Kalmár 1957).

The functions g of h that depend upon only a finite number of coord nates of h play an important role in mathematics and will here be called functions of *finite structure*. For many purposes, however, this family is too small. In particular, as has just been seen, the real-valued functions of finite structure are inadequate as the domain of definition of the integral σ.

Countably additive experience brings to mind the smallest family of functions that contains the real-valued functions of finite structure and is closed under pointwise limits; but this family is ill suited to the more general finitely additive theory of integration. The present section discusses a different natural family of functions that is much richer than the family of finite structure and is well suited to finitely additive integration.

For each mapping g of H into any set R, gf is, as in Section 6, the mapping of H into R defined by $gf(h) = g(fh)$.

Call a family G of mappings of H into a set R *finitarily closed* if $g \in G$ whenever $gf \in G$ for all f in F. Call the smallest finitarily closed family G of mappings of H into R that contains the constant functions the *finitary* mappings. To put it a little differently, g is a finitary mapping of H into R if every finitarily closed family of mappings of H into R that contains the constant mappings contains g. Several useful characterizations of finitary mappings will now be developed.

A mapping g of H into R has *structure* 0 if g is a constant. For each ordinal α greater than 0, g has *structure at most* α if, for each f in F, there is an ordinal β less than α for which gf has structure at most β.

If, for some α, g has structure at most α, g is *structured*. If g is structured, let $\alpha(g)$ be the first ordinal α for which g has structure at most α, and call $\alpha(g)$ the *structure* of g. Clearly, $\alpha(g) = [\sup \alpha(gf)] + 1$ or $\sup \alpha(gf)$ according as $\sup \alpha(gf)$ is or is not attained or, equivalently, according as $\alpha(g)$ is or is not the successor of an ordinal.

If F contains only one element, every g is of structure 0. If F contains a finite number of elements, but more than one, there is a g of every finite structure α but no g of infinite structure. If F is infinite, any α that does not exceed F in cardinality can occur as a structure, but no others can.

It was justifiable to call a g that depends on only the first n coordinates of h of finite structure, as we did, since such a g clearly has structure at most n for the finite ordinal n.

When should a mapping of H into R be called continuous? Since F and R are introduced here simply as abstract sets, the discrete topology (the topology in which every subset is open) is the only topology that suggests itself for them. But for H it is natural to introduce the Cartesian-product topology corresponding to the discrete copies of F of which H is the Cartesian product. That is, a subset S of H is *open* if, and only if, for every history h in S, there is an n such that every history h' which agrees with h in the first n coordinates is also in S. A mapping g of H into R is, then, *continuous* if and only if $g^{-1}(A)$ is an open subset of H for every subset A of R. Equivalently, g is continuous if and only if $g^{-1}(A)$ is both an open and a closed subset of H for every subset A of R.

A mapping g of H into R is *determined* if, for every h, there is an n such that $g(h') = g(h)$ for every history h' that agrees with h in the first n coordinates. Let $n_g(h)$ be the smallest such n, possibly 0, and say that g is *determined at (time)* n_g. Clearly, n_g is itself a determined function and is determined no later than g, that is, $n_{n_g} \leq n_g$.

We are indebted to conversation with John Myhill for the topological aspect of the following characterization theorem. This aspect is the point of departure for (Kalmár 1957).

THEOREM 1. The following four conditions on a mapping g of H into R are equivalent.

(*a*) g is finitary.

(*b*) g is structured.

(*c*) g is continuous.

(*d*) g is determined.

Proof:

(a) implies (b): The set of structured g's contains the constants and is finitarily closed.

(b) implies (c): The functions of structure 0, the constants, are certainly continuous. Suppose that all functions of structure less than α are continuous, and consider a g of structure α. Then gf is continuous for each f. That is, for each f and h, there is an n such that, if h' agrees with h through the first n coordinates, then $gf(h) = gf(h')$. Therefore $g(fh) = g(fh')$, where fh' is any history that agrees with fh in the first $n + 1$ coordinates. Thus g is continuous.

(c) implies (d): Obvious.

(d) implies (a): Suppose that some determined g is not finitary. Then, for some f_1, gf_1 is not finitary, and so, for some f_2, gf_1f_2 is not finitary. This generates an infinite sequence $h = (f_1, f_2, \cdots)$ such that $gf_1 \cdots f_n$ is not finitary for any n. Then for no n is $gf_1 \cdots f_n$ constant, contradictory to the assumption that g is determined. ♦

A set of histories is a *finitary event* if the indicator of the set is a finitary function. (The *indicator* of a set is, as usual, the function that is 1 on the set and 0 off the set.)

COROLLARY 1. An event is finitary if and only if it is a closed and open subset of H.

(In the study of sets of sequences of integers it is usual to construct various hierarchies beginning with the closed, open sets, such as Baire's classification of the Borel sets (Sierpinski 1950), (Liapunov, Schegolkow, and Arsenin 1955), and (Addison 1958). As early as 1930, Lusin (1930, p. 85) classified the closed, open subsets of the irrationals into a hierarchy beginning with the null set and the set of all irrationals similar, if not isomorphic, to the hierarchy of closed, open sets of sequences implicit in (Kalmár 1957).)

If F is a finite set, the finitary functions are relatively limited in structure.

THEOREM 2. If F is finite and g is finitary, then g is of finite structure and assumes only a finite number of values.

Proof: Immediate from the definition of finitary. ♦

Finitary or, equivalently, determined functions can be very freely combined.

THEOREM 3. If $g_1, \cdots, g_k$ are determined, the k-tuple $(g_1, \cdots, g_k)$ is itself a determined function, determined at the maximum of the times at which the g_i are determined.

THEOREM 4. Any function of a determined function g is determined no later than g.

8. INDUCTIVELY INTEGRABLE FUNCTIONS

A bounded, real-valued, finitary function is called *inductively integrable.* The next theorem partially justifies this term and fulfills the program of definition of the integral σ begun in Section 6.

THEOREM 1. There is one and only one function E defined on all pairs (σ, g), where σ is a strategy and g is an inductively integrable function, such that $E(\sigma, c) = c$ for every constant c and

$$E(\sigma, g) = \int E(\sigma[f_1], gf_1)\, d\sigma_0(f_1). \tag{1}$$

Proof: An E will be defined inductively, checked, and shown to be unique.

If g is of structure 0, it is a constant c. Let $E(\sigma, c) = c$. If E has been so defined for all σ and for all bounded g of structure less than α that $|E(\sigma, g)| \leq \sup |g|$, let $E(\sigma, g)$ be defined for a g of structure α by (1). This is an unambiguous definition covering g's of every structure α and, therefore, all inductively integrable g's. So defined, $E(\sigma, g)$ plainly satisfies the condition of the theorem. Finally, it is easy to see by induction that no other function does. ♦

Usually, $E(\sigma, g)$ will be abbreviated to σg. This is in harmony with the definition (6.1) already given for functions of finite structure, and the integral σ has the linearity and positivity appropriate to a probability integral. These facts are easily checked, mainly by induction, and are here recorded.

THEOREM 2. If σ is a strategy, g and g' are inductively integrable, and c and c' are constants, then:

(*a*) If $g = c$, $\sigma g = c$.

(*b*) If g is of finite structure, (6.1) applies.

(*c*) (6.3) applies.

(*d*) $cg + c'g'$ is inductively integrable.

(*e*) $\sigma(cg + c'g') = c\sigma g + c'\sigma g'$.

(*f*) If $g \geq 0$, $\sigma g \geq 0$.

For the main purposes of this book, σg has now been defined for an adequate domain of g's, but as with any nonnegative measure, the Eudoxus extension can be defined and is likely to prove useful. Any bounded function g of histories exceeds some inductively integrable functions and is exceeded by others. Let $E^*(\sigma, g)$ and $E_*(\sigma, g)$ be the upper and lower integrals of g with respect to σ.

If, for some σ and g, $E^*(\sigma, g)$ and $E_*(\sigma, g)$ are equal, g is *integrable* under σ, and the domain of definition of E can then be extended by letting $E(\sigma, g)$, or its abbreviation σg, be $E_*(\sigma, g)$. When it is desired to call attention to the possibility that an integrable g is not inductively integrable, g is called *Eudoxus integrable* under σ in harmony with a definition in Section 3.

THEOREM 3. If gf is (Eudoxus) integrable under $\sigma[f]$ for every f and if g is bounded, then g is (Eudoxus) integrable under σ and (1) holds.

The import of Theorem 3 is that (after Eudoxus extension) E cannot be extended to more bounded functions by (1). The converse of Theorem 3 does not hold. But a strengthened form of Theorem 3 has a converse, as the next theorem, preceded by a lemma, asserts.

LEMMA 1. For all strategies σ and all bounded functions g,

$$E^*(\sigma, g) = \int E^*(\sigma[f], gf)\, d\sigma_0(f), \tag{2}$$

and

$$E_*(\sigma, g) = \int E_*(\sigma[f], gf)\, d\sigma_0(f). \tag{3}$$

THEOREM 4. A bounded function g is integrable under σ if and only if

$$\int E^*(\sigma[f], gf)\, d\sigma_0(f) = \int E_*(\sigma[f], gf)\, d\sigma_0(f). \tag{4}$$

In this event, $E(\sigma, g)$ equals (4).

As with any probability integral, σ can be viewed as a probability measure defined on those subsets M of H that have integrable indicator functions; σM, the probability of M under σ, is defined as the integral with respect to σ of the indicator function of M. If M is finitary, the indicator of M is even inductively integrable. When it seems desirable to note that the Eudoxus integral is invoked, σM is called the *Eudoxus measure* of M.

(In addition to the general technique of extending any integral to some unbounded functions in terms of the behavior of their truncations (as will be reviewed and used later), a special technique suggests itself in the case of strategies. For example, if gf is integrable under $\sigma[f]$ for each f, and if $E(\sigma[f], gf)$, as a function of f, is integrable with respect to σ_0, then (1) suggests the value for $E(\sigma, g)$, even if g is unbounded, etc.)

The next theorem and its corollary help to bring out the meaning of integration with respect to σ, though they are not applied in this book. The theorem is not difficult to prove, especially with techniques in the next section and in Section 3.2.

THEOREM 5. For a bounded g, the following conditions are equivalent:

(*a*) g is Eudoxus integrable under every strategy σ.

(*b*) For every history h and positive ϵ, there is an n such that if h' agrees with h at the first n coordinates then $|g(h) - g(h')| \leq \epsilon$.

(*b'*) g is continuous from the product topology on H to the ordinary topology on the real numbers.

(*c*) For every positive ϵ there is an inductively integrable g' such that $|g - g'| \leq \epsilon$.

(*c'*) g is the uniform limit of a sequence of inductively integrable functions.

(*d*) gf is Eudoxus integrable under every σ for all f.

In the light of Theorem 5, Theorem 2 is valid for the class of functions that are integrable under all σ.

COROLLARY 1. M has a Eudoxus measure for all strategies σ if and only if M is a finitary event.

Countably additive counterparts of ideas from this and the preceding two sections have been explored by J. P. Raoult (1964*a*).

9. STOP RULES, POLICIES, AND TERMINAL FORTUNES

As we define it, the policy of a gambler is not specified by σ alone. There must also be a stop rule, a positive integer $t(h)$ assigned to each history h, expressing the gambler's decision to terminate play after the $t(h)$th move and to accept $f_{t(h)}$ as his terminal fortune. Stop rules are familiar in the theory of games, statistics, and the theory of probability. Formally, a *stop rule* is a mapping t from H to the positive integers such that, if h' agrees with h through the first $t(h)$ coordinates, then $t(h') = t(h)$. And a *policy* π is the pair (σ, t); π is *available in* Γ *at* f if σ is.

This definition of policy does not cover the possibility that the gambler will elect to settle for his initial fortune without gambling at all. Expressing this liberty in terms of a policy has always led us into technical annoyances. However, the gambler can ordinarily produce the effect of terminating without gambling by letting his first gamble σ_0 be the one-point measure $\delta(f_0)$ and letting t be identically 1.

A stop rule is determined in the sense of Section 7; in fact, t is evidently determined by time t at the latest. It is therefore finitary and continuous, and has a structure $\alpha(t)$, according to Theorem 7.1.

The following is essentially a famous and simple theorem of König (1936, Theorem 6, p. 81).

THEOREM 1. If F is finite, every stop rule is bounded.

Proof: Since every stop rule t is determined, Theorem 7.1 applies to show that it is finitary. Theorem 7.2 then implies that it is bounded. ◆

It may seem too strict to require the gambler to terminate every history. Would it not be enough to require him to terminate almost certainly? This question will be explored as soon as is practical (in

Section 11), when almost-terminating policies will be shown to offer no material advantage to the gambler.

Associated with each stop rule t is an important mapping of H into F, the *terminal fortune* f_t, which is, for each h, the $t(h)$th coordinate of h; that is $f_t(h) = f_{t(h)}$.

Had we insisted on allowing termination before the first gamble, that is, allowing t to be 0, then f_t would, in such a case, be undefined. To define it as f_0 and reckon f_0 as part of the history replaces the concept of a strategy, in effect, by that of a family of strategies and eliminates a small technical annoyance at the price of a larger one, as would leaving f_t undefined when t is 0.

If every coordinate of h is f, then $f_t = f$; so f_t takes all values in F.

THEOREM 2. The function f_t is determined. Aside from the uninteresting possibility that F contains but one element, f_t is determined exactly at time t; that is, $n_{f_t} = t$.

Intuitively: There is no forecasting with certainty, at any time before t, what fortune the gambler will possess when he stops. By contrast, the time t of stopping is sometimes decided before stopping occurs. For example, $t(h)$ might be 25 for all h so that the structure of t would be 0.

Proofs by induction often involve functions f_tf derived from f_t. In accordance with Section 7, $f_tf(h) = f_t(fh)$ for all h. Since t is a stop rule, either $tf \equiv 1$ or else tf is never 1. In the latter case, $tf - 1$ is itself a stop rule $t[f]$. As is easy to verify,

$$f_tf \equiv f \qquad \text{if } tf \equiv 1 \tag{1}$$
$$= f_{t[f]} \qquad \text{otherwise.}$$

Intuitively: The gambler finds himself at f after his first gamble σ_0; he stops then and there if $tf \equiv 1$; otherwise he stops later—when and where being governed by a new stop rule $t[f] = tf - 1$. Sometimes the notation $t[f]$ is used for $tf - 1$, even when $tf - 1$ is 0 and, therefore, technically not a stop rule.

THEOREM 3. If $t[f]$ is a stop rule and F has at least two elements, then the structure of $f_{t[f]}$ is strictly less than that of f_t. For any t, the structure of f_t is at least that of t.

If a gambler follows a policy $\pi = (\sigma, t)$ until it terminates after some partial history $p = (f_1, \cdots, f_n)$ at the fortune f_n, and if he then resumes gambling according to a policy $\pi'(p)$, he is, in effect, adopting a composite policy π''. With respect to π'', $\pi'(p)$ represents the conditional distribution of the behavior of the gambler given that he has experienced p up to time t. The next three paragraphs, which are used only occasionally and can be skimmed or deferred, present this somewhat technical notion formally.

Let $h^n = (f_{n+1}, f_{n+2}, \cdots)$ denote the part of the history h after time n. Thus $h^{t(h)} = (f_{t(h)+1}, f_{t(h)+2}, \cdots)$ is the part of h after $t(h)$. Similarly, let $p_t(h) = (f_1, \cdots, f_{t(h)})$ be the part of h up to and including $t(h)$. The composition of a stop rule t with a family of stop rules $t'(p)$, one for each partial history p, means the stop rule t'' that continues beyond t according to $t'(p_t(h))$;

$$t''(h) = t(h) + t'(p_t(h))(h^{t(h)}). \tag{2}$$

There is no fundamental reason why $t''(h)$ should not be allowed to equal $t(h)$ for some, or all, histories h, contrary to the convention adopted earlier according to which $t'(p)$ cannot be 0. But the additional freedom would ordinarily be of no material advantage and would entail some additional complication.

Consider a gambler who pursues a policy $\pi = (\sigma, t)$ until it terminates after a partial history p and who then continues according to a family of strategies $\sigma'(p)$. In effect, he composes the strategy $\sigma'' = (\sigma_0'', \sigma_1'', \cdots)$, where

$$\sigma_n''(f_1, \cdots, f_n) = \begin{cases} \sigma_n(f_1, \cdots, f_n) & \text{for } n < t(h), \\ \sigma_0'(p_t(h)) & \text{for } n = t(h), \\ \sigma_{n-t(h)}'(p_t(h))(f_{t(h)+1}, \cdots, f_n) & \text{for } n > t(h). \end{cases} \tag{3}$$

Formally, *the policy π followed by the family of policies $\pi'(p)$* is the *composite policy $\pi'' = (\sigma'', t'')$*.

The policy $\pi = (\sigma, t)$ is largely tantamount to the distribution of f_t under σ, that is, to the gamble $\gamma(\pi)$ for which $\gamma(\pi)E$ is, for each subset E of F, the probability under σ that $f_t \in E$. In terms of $\gamma(\pi)$, the *gamble induced by π*,

$$\begin{aligned} \gamma(\pi)v = \gamma(\sigma, t)v &= \int v(f)\, d\gamma(\pi)(f) \\ &= \int v(f_t(h))\, d\sigma(h) \end{aligned} \tag{4}$$

for any bounded function v of fortunes. Of course $v(f_t)$ is a finitary function of h (Theorems 2 and 7.4) and is therefore inductively integrable.

Just as a policy π induces a distribution $\gamma(\pi)$ on fortunes, defined in terms of the mapping h into $f_t(h) = f_{t(h)}$, it also induces a distribution $\eta(\pi)$ on partial histories p, defined in terms of the mapping of h into $p_t(h)$. Since p_t is a determined function to partial histories, $w(p_t(h))$ is inductively integrable for all bounded functions w of partial histories, and

$$\eta(\pi)w = \eta(\sigma, t)w = \int w(p)\, d\eta(\pi)(p) \tag{5}$$
$$= \int w(p_t(h))\, d\sigma(h).$$

If π'' is the policy π followed by the family of policies $\pi'(p)$, then

$$\gamma(\pi'')v = \int \gamma(\pi'(p))v\, d\eta(\pi)(p), \tag{6}$$

as is not hard to verify.

In some applications, $\pi'(p)$ depends only on the final element f_n, say f, of p. Then $\pi'(p)$ can advantageously be written $\pi'(f)$, and (6) specializes essentially to

$$\gamma(\pi'')v = \int \gamma(\pi'(f))v\, d\gamma(\pi)(f). \tag{7}$$

A brief and light digression on the generation of new stop rules from old ones concludes this section. If t is a stop rule and ϕ maps the positive integers into themselves, then $\phi(t)$ may or may not also be a stop rule. For example, t^2 is a stop rule, but the largest integer not exceeding $t^{1/2}$ typically is not. Which are the ϕ that transform all stop rules into stop rules? Among them are certainly those for which $\phi(i) \geq i$ for all positive integers i. The only others are those which, for some positive k, satisfy

$$\begin{aligned} \phi(i) &\geq i \qquad \text{for } i \leq k \\ &= k \qquad \text{for } i > k. \end{aligned} \tag{8}$$

With the convention that k can be ∞, (8) characterizes the ϕ that carry stop rules into stop rules. What integral-valued functions ψ of two integral variables carry each pair of stop rules into a stop rule? Just those for which $\psi(i, j)$ satisfies the condition (8) in i for each j and in j for each i. Examples are: min (i, j), max (i, j), $i + j$, and ij. The principle extends immediately from two to any number of arguments, finite or infinite.

10. THE UTILITY OF FORTUNES, POLICIES, AND HOUSES

How are the gambler's desires or aspirations to be described mathematically? For most of the concrete problems in this book, the description is very simple: Some fortunes are satisfactory and others are not; the gambler seeks to maximize the probability of terminating play at a fortune that is satisfactory. With no extra trouble, and with some advantage, a more general and realistic model of the gambler's desires is adopted. Namely, it is assumed, in accordance with much current theory about behavior in the face of uncertainty, that the gambler's desires are summarized, in a certain sense, by a function u from fortunes to real numbers, called the *utility function*. The worth or utility of the fortune f to the gambler is $u(f)$. If the gambler's only objective is to cease gambling with a numerical fortune as large as f' if possible, then $u(f)$ can be taken to be 0 or 1 according as f is or is not exceeded by f'.

This book is not confined to such two-valued utility functions, because the general function seems natural to the theory and proves to be a useful tool in the study of two-valued utilities. (For background about the theory of utility see Chapter 5 of (Savage 1954) and references cited there.) In order to avoid distracting complications we shall, with casual exceptions in Chapter 12, assume u to be bounded.

In many mathematical studies of gambling and insurance, the "gambler" is interested in maximizing his expected cash, in which case cash—which is not necessarily bounded—is identified with utility. This utility function is often tacit when it is said that gambling in a commercial gambling house does not pay.

According to the intended interpretation of utility, the gambler will value a policy π according to the expected value under π of the utility of its terminal fortune f_t. Thus, if $\pi = (\sigma, t)$ is a policy, it is appropriate to call

$$u(\pi) = u(\sigma, t) = \sigma u(f_t) \tag{1}$$

the *utility* of π to the gambler. If $u(f)$ is simply 1 or 0 according as f is or is not satisfactory, then the utility of π is simply the probability under σ that f_t is satisfactory—the probability under π that the gambler will succeed.

With the policy $\pi = (\sigma, t)$ is associated the family of couples $\pi[f] = (\sigma[f], t[f])$. (The meaning of the components is as in Sections 5 and 9.) Except where $t[f] \equiv 0$, $\pi[f]$ is itself a policy, the *conditional policy given* f, and it has a utility $u(\pi[f])$; where $t[f] \equiv 0$, it is useful to let $u(\pi[f])$ be $u(f)$. In these terms,

$$u(\pi) = \int u(\pi[f])\, d\sigma_0(f); \tag{2}$$

the utility of a policy is the expectation under its initial gamble of the utility of the part of the policy that remains after the first gamble, so to speak.

The gambler's object is to make $u(\pi)$ as large as possible subject to his initial fortune f and the rules of the gambling house Γ or, if he finds that preferable, to settle for $u(f)$. Therefore, let

$$U(f) = \text{the maximum of } \sup u(\pi) \text{ and } u(f), \tag{3}$$

confining the supremum to those policies π that are available in Γ at f, and call the number $U(f)$ the *utility* of Γ at f. To make explicit the dependence of U on Γ and u, it is sometimes better to write Γu for U.

For the purposes of most of this book, the general "gambler's problem" is to find U when Γ and u are given and to determine policies that are optimal or nearly optimal. Simple though the concept of U defined by (3) is, few passages in this chapter thus far have had any other objective than to contribute to the definition. In our experience, this formulation of the utility of Γ has been fruitful and satisfying. The next section shows that a slightly different approach leads to the same U. The next chapter enlarges the notion of gambler's problem by introducing a function V never greater than U and typically equal to U, that seems to have a more fundamental claim to measuring the worth of Γ.

It is implicit in the extramathematical interpretations that have led to the definition of U that there is no rule restricting the gambler from terminating play at any stage and enjoying the utility of the fortune at which he then finds himself. A foundation for the study of problems with such rules is given in the next chapter, and some further suggestions are in Chapter 12.

Relaxation of the rules or an increase in the utility function is never unwelcome to the gambler.

THEOREM 1. If $\Gamma \subset \Gamma'$ and $u \leq u'$ (that is, $\Gamma(f) \subset \Gamma'(f)$ and $u(f) \leq u'(f)$ for all f), then $\Gamma u \leq \Gamma' u'$.

The mapping of u into Γu is not only nondecreasing in u, as Theorem 1 implies; it is also strongly continuous in u, as the following asserts.

THEOREM 2. If u_n converges uniformly to u, then Γu_n converges uniformly to Γu. In fact, if $|u - u'| \leq \epsilon$, then $|\Gamma u - \Gamma u'| \leq \epsilon$.

11. INCOMPLETE STOP RULES

A rule that assigns a terminal fortune to almost all histories, though perhaps not absolutely all, claims some attention. This section shows that the gambler can gain no material advantage from incomplete stop rules that terminate for almost all histories.

An *incomplete stop rule* is a function t defined on histories and taking as values positive integers, and possibly ∞, such that if $t(h) < \infty$ and h' agrees with h through the $t(h)$th coordinate then $t(h) = t(h')$.

Example 1. Let A be a subset of F, and let $t(h)$ be the first n, if any, for which $f_n \in A$; for all other h, $t(h) = \infty$.

Suppose t is incomplete but the set of histories h for which $t(h) < \infty$ has Eudoxus probability 1 under a strategy σ. Then plainly, no matter how $u(f_t)$ is extended to be a bounded function defined for all h, $u(f_t)$ is Eudoxus integrable under σ, and $\sigma u(f_t)$ has the same value for all such extensions; this value is designated by $u(\sigma, t)$ in harmony with (10.1).

THEOREM 1. If t is an incomplete stop rule and T is a finitary (or even closed) subset of $\{h: t(h) < \infty\}$, then there is a complete stop rule t' that agrees with t for all $h \in T$.

Proof: Let h be any history. If there is an h' in T that agrees with h in their first $t(h')$ coordinates, let $t'(h) = t(h')$. If no such h' exists, let $t'(h)$ be the least n such that no history h' that agrees with h in their first n coordinates is in T. That t' is a stop rule is easily verified. ◆

THEOREM 2. If an incomplete stop rule t is finite with σ-probability 1, then there is a complete stop rule t' for which $|\sigma u(f_t) - \sigma u(f_{t'})|$ is arbitrarily small.

Proof: Theorem 1 applies. ◆

The decision to restrict attention to stop rules as originally defined therefore seems innocuous. In particular, the important function U is not affected by this decision.

The incomplete stop rules, as functions from H to the positive integers and ∞, have the natural partial ordering in which $t \leq t'$ if and only if $t(h) \leq t'(h)$ for all h.

THEOREM 3. Under the natural partial ordering, the stop rules form a sublattice of the lattice of incomplete stop rules. The maximum of any finite set, as well as the minimum of any set, of stop rules (incomplete stop rules) is a stop rule (incomplete stop rule). The minimum of a stop rule and an incomplete stop rule is a stop rule.

The concluding paragraph of Section 9 can be so extended for incomplete stop rules as to embrace Theorem 3, which is also easy to check directly.

12. FUNCTIONS THAT MAJORIZE U

This section revolves about a simple theorem which often applies to show that some function Q is at least as large as U. We are much indebted to Aryeh Dvoretzky for bringing home to us the truth and importance of the theorem.

If $\gamma Q \leq Q(f)$ for some fortune f, gamble γ, and function Q on F, then, consonant with certain publications on Markov processes such as (Hunt 1957): Q is *excessive* for the gamble γ at f. If Q is excessive for σ_0 at f and for $\sigma_n(f_1, \cdots, f_n)$ at f_n for all n and $(f_1, \cdots, f_n)$, then Q is *excessive* for the strategy σ, and for each policy (σ, t), at f. If, for all f, Q is excessive at f for each γ in $\Gamma(f)$, then Q is *excessive* for the gambling house Γ.

Plainly, Q is excessive for Γ if and only if, for all f, Q is excessive for every policy (strategy) available in Γ at f. Also, Q is excessive for a policy (strategy) at f if and only if that policy (strategy) is available at f in some house for which Q is excessive.

Here is one more expression of the conservation-of-subfairness idea, whose importance in probability was mentioned in Section 1.3.

LEMMA 1. If Q is excessive for π at some f', then $Q(\pi) \leq Q(f')$.

Proof: The following calculation is justified by (10.2), Theorem 9.3, and an induction on the structure $\alpha(f_t)$.

$$Q(\pi) = \int Q(\pi[f])\, d\sigma_0(f) \tag{1}$$

$$\leq \int Q(f)\, d\sigma_0(f)$$

$$\leq Q(f'). \quad \blacklozenge$$

THEOREM 1. If $u \leq Q$ and Q is excessive for Γ, then $U \leq Q$. (That is, if $u(f) \leq Q(f)$ and $\gamma Q \leq Q(f)$ for all f and all γ in $\Gamma(f)$, then $U(f) \leq Q(f)$ for all f.)

Proof: Since $u(\pi) \leq Q(\pi)$ for all π, Lemma 1 implies that $u(\pi) \leq Q(f)$ for all f and all π in Γ at f. Consequently, $U \leq Q$. ◆

This theorem has previously been used implicitly in gambling problems, in particular by Blackwell (1954) and by de Finetti (1939, 1940), who traces the idea through (Bertrand 1907) to De Moivre. Other well-known problems whose solutions implicitly involved this theorem are described in Sections 12.4 through 12.9. Throughout this book, Theorem 1 will play an explicit and important role.

Though Theorem 1 has been easy to prove directly, a slightly longer and more general-looking path shows its relation to some current probability theory. Consider a strategy σ, a constant Q_0, and a uniformly bounded sequence of functions $Q_1, Q_2, \cdots$ defined on H and such that:

(*a*) Q_n depends only on the first n coordinates $f_1, \cdots, f_n$ of h.

(*b*) The expectation under σ of Q_{n+1} given $f_1, \cdots, f_n$ is at most Q_n.

Such a sequence Q_n ought, in the spirit of current usage (Doob 1953), to be called a lower semimartingale for the increasing pattern of information $f_1, \cdots, f_n$. The condition (*b*) is of course interpreted formally thus:

$$(2) \quad \int Q_{n+1}(f_1, \cdots, f_n, f_{n+1})\, d\sigma_n(f_1, \cdots, f_n)(f_{n+1}) \leq Q_n(f_1, \cdots, f_n).$$

The following theorem is a finitely additive version of a prominent theorem about semimartingales; compare (Doob 1953, Theorem VII2.2).

THEOREM 2. If $\{Q_n\}$ is a lower semimartingale with respect to σ, and t is a stop rule, then $\sigma Q_t \leq Q_0$.

To paraphrase, the expected terminal cash fortune yielded by any stop rule applied to a succession of uniformly bounded fortunes resulting from subfair bets is at most the initial fortune.

The proof of Theorem 2 is sufficiently suggested by that of Lemma 1.

With Theorem 2, Lemma 1 can be re-proved thus. Let $Q_0 = Q(f)$, $Q_n(f_1, \cdots, f_n) = Q(f_n)$, where Q is as in Lemma 1. Then $\{Q_n\}$ is plainly a lower semimartingale with respect to any σ in Γ at f. Theorem 2 then applies to yield the conclusion of Lemma 1 again.

(Theorem 2 can be viewed as a special case of Lemma 1 by means of a device that will be formalized in Section 12.2, namely, the device of regarding partial histories of a given space of fortunes F as the fortunes of a new space of fortunes.)

If $U(f)$ were to be defined with reference to sup $u(\pi)$ over some restricted class of policies, for example by contracting Γ or by imposing measurability conditions, $U(f)$ would not be increased thereby, and so Theorem 1 would remain true *a fortiori*.

Theorem 1 gives a useful sufficient condition for a function Q to *majorize* U, that is, to be at least as large as U everywhere. Still more, if for some f there is a π available in Γ at f for which $u(\pi)$ equals or is arbitrarily close to $Q(f)$, then clearly, for that f, $Q(f)$ must actually be $U(f)$, as the following theorem underlines.

THEOREM 3. If $Q(f) \leq U(f)$ for some f, Q is excessive for Γ, and Q majorizes u, then $Q(f) = U(f)$. Consequently, if the first hypothesis holds for all f, then Q is U.

The remainder of this section sometimes applies to show that U is at least as large as some function Q. This variant of Theorem 1 can be skipped or postponed.

A bounded function Q on F is *deficient* at f for a strategy σ if: (*a*) $\sigma_0 Q \geq Q(f)$ and (*b*) for all $n \geq 1$ and all $f_1, \cdots, f_n$,

$$\sigma_n(f_1, \cdots, f_n)Q \geq Q(f_n).$$

THEOREM 4. Suppose that for some f there is a σ available in Γ at f and a Q deficient at f for σ, and that for every positive ϵ there is a stop rule t for which $u(f_t) \geq Q(f_t) - \epsilon$ with σ-probability at least $1 - \epsilon$. Then $U(f) \geq Q(f)$.

Proof: $U(f) \geq \sigma u(f_t) \geq \sigma Q(f_t) - \epsilon - \epsilon \sup (Q - u) \geq Q(f) - \epsilon - \epsilon \sup (Q - u)$, where the proof that $\sigma Q(f_t) \geq Q(f)$ is simple and imitates that of Lemma 1. ◆

13. DIGRESSION ABOUT COUNTABLE ADDITIVITY

Since noncountably additive gambles are often regarded as pathological, some readers may now be interested in conditions guaranteeing that the gambler's "earning ability" is not affected if he is restricted to conventional gambles and strategies. Others may wish to omit this section, or to postpone it until the same theme is touched upon later.

Theorem 12.3 can be useful in verifying that some new restriction on the gambler does not affect his earning ability. One application is this. It may well happen that in a first formulation of the problem the gambles are defined only on some class (perhaps a Boolean algebra or sigma-field) of special interest. Perhaps, too, they are countably additive when first defined. There may well be hope that it is immaterial how the definition of the gambles is extended beyond this algebra. For example, the γ's might originally be defined on the Borel sets of the real line, u might be Borel measurable, and, if the problem is happily formulated, there will be reason to hope that there is some policy arbitrarily nearly optimal among the policies that deserve to be called Borel measurable.

A definition of measurability for policies suggests itself: With respect to a sigma-field of subsets of F called measurable, σ is *measurable* if $\sigma_n A$ is measurable in its arguments $f_1, \cdots, f_n$ for every measurable set A of fortunes. And $\pi = (\sigma, t)$ is *measurable* if σ is measurable in the sense just defined and if t is measurable in the ordinary sense for Cartesian products of measurable spaces. If u and π are measurable, the value of $u(\pi)$ is determined by the values that the gambles take on measurable sets of fortunes, as may be seen by induction on the structure of t.

Investigation will show (in Section 16) that, if $\Gamma(f)$ and $u(f)$ depend sufficiently smoothly on f, then measurable, nearly optimal strategies exist and U is measurable. Sometimes, even where $\Gamma(f)$ does not depend smoothly on f and no useful measurable strategies exist, the values of the gambles γ for nonmeasurable sets are nonetheless immaterial and U is measurable. The values of γ for nonmeasurable sets may, however, become quite material for U when such a problem is modified by placing a bound on the time allowed for gambling.

Example 1. Let F be the closed unit interval $[0, 1]$ together with $\{2\}$, and let K, with indicator function k, be a subset of $[0, 1]$ with inner (Lebesgue) measure 0 and outer measure 1. For f in K, let $\Gamma(f)$ consist of $\delta(f)$ and $\delta(2)$; for f not in K, let $\Gamma(f)$ consist of $\delta(f)$ and γ, where γ has these three compatible properties: $\gamma\{2\} = \frac{1}{2}$; γA is half the Lebesgue measure of A if A is a measurable subset of $[0, 1]$; $\gamma K = \frac{1}{2} - \alpha$ for an arbitrary α in $(0, \frac{1}{2}]$. Finally, suppose $u(f) = 0$ for $f \leq 1$ and $u(2) = 1$. The supremum of $u(\pi)$ over policies π that begin at $f \in [0, 1]$ and are such that $t \leq n$ is easily seen to be $(1 - \frac{1}{2}\alpha^{n-1}) + \frac{1}{2}\alpha^{n-1}k(f)$. This depends on α, the probability under γ of a nonmeasurable set (and is not measurable in f). The fact that the constrained supremum approaches 1 in n and that $u \leq 1$ shows that $U = 1$. Consequently, U is independent of α and measurable in f. In this example, $u(\pi) \leq \frac{1}{2}$ for any measurable π beginning at f in $[0, 1]$ and not in K.

Let $Q(f)$ be the maximum of $u(f)$ and the supremum of $u(\pi)$, where π is restricted to, say, Borel-measurable policies in Γ at f, and suppose, for a moment, that Q is Borel measurable. Theorem 12.3 implies that the manner of extending the γ's is immaterial if $\gamma Q \leq Q(f)$ for all f and γ in $\Gamma(f)$—a condition which is, of course, verifiable in terms of the unextended γ's alone. A later conclusion (Theorem 14.1) implies that this sufficient condition for Q to be U is also necessary. The principle is frequently illustrated in later chapters where a Q comes under consideration because it is approached by means of Borel-measurable (or even much more regular) policies that are suspected of being optimal or nearly so. Clearly, for such a Q, $Q(f) \leq U(f)$. Verification that $\gamma Q \leq Q(f)$ and $u(f) \leq Q(f)$ for all f and γ in $\Gamma(f)$ then shows that Q is indeed U and that "regular" policies can be nearly optimal. If Q is not measurable—and this could easily happen for a sufficiently ill-formulated problem—then γ should be replaced by its upper integral γ^* in this test.

14. U IS EXCESSIVE FOR Γ.

The first hypothesis of Theorem 12.1 is obviously automatic for every majorant Q of U. The second, namely that Q be excessive for Γ, is not automatic. To see that it is not, add a large constant to any function Q that is not excessive for Γ. For U itself, however, even the second hypothesis is automatic:

THEOREM 1. $\gamma U \leq U(f)$ for all f and all γ in $\Gamma(f)$.

Proof: Suppose that, for some f' and γ in $\Gamma(f')$, $\gamma U = U(f') + \delta$, with $\delta > 0$. Let $\pi(f) = (\sigma(f), t(f))$ be, for each f at which $u(f) < U(f)$, a policy in Γ at f for which $u(\pi(f)) \geq U(f) - \frac{1}{2}\delta$. Construct any new policy π' that begins with γ and terminates at time 1 if $u(f_1) = U(f_1)$ and otherwise continues with $\pi(f_1)$. More formally, $\sigma_0' = \gamma$; $\sigma'[f]$ is arbitrary if $u(f) = U(f)$ and is $\sigma(f)$ otherwise; and

$t'(fh)$ is 1 if $u(f) = U(f)$ and is $t(f)(h) + 1$ otherwise.

$$
\begin{aligned}
(1) \qquad U(f') &\geq u(\pi') \\
&= \int u(\pi'[f])\, d\sigma_0'(f) \\
&= \int u(\pi'[f])\, d\gamma(f) \\
&\geq \int \left(U(f) - \frac{\delta}{2}\right) d\gamma(f) \\
&= U(f') + \frac{\delta}{2}. \quad \blacklozenge
\end{aligned}
$$

As Theorem 1 says, U is excessive for Γ. In dynamic programming applications, this fact is called "the principle of optimality". However, in a conventional countably additive theory, where $U(f)$ is ordinarily the supremum of $u(\pi)$ for measurable π in Γ at f, Theorem 1 would pose the difficulty that an optimal or near optimal $\pi(f)$ might not vary measurably with f, so the integral in (1) might not exist. In fact, the theorem is not true for the U associated with measurable policies; it obviously fails for Example 13.1.

COROLLARY 1. U is the smallest of all functions that are excessive for Γ and majorize u. Consequently, $Q = U$ if and only if $u \leq Q \leq U$ and Q is excessive for Γ.

Of course, this corollary which characterizes U subsumes Theorem 1 and Theorems 12.1 and 12.3.

COROLLARY 2. If $u \leq v \leq U$, then $\Gamma v = U$. In particular, $\Gamma U = U$.

If, for all f, $\gamma \in \Gamma(f)$ if and only if $\gamma \in \Gamma'(f)$ or $\gamma \in \Gamma''(f)$, then Γ is the *union* of Γ' and Γ''. Of course, U is then at least as large as the maximum of U' and U''.

THEOREM 2. Let Γ be the union of Γ' and Γ''. Then U' is excessive for Γ'' if and only if $U = U'$. In this event, $U' \geq U''$.

15. LIMITED PLAYING TIME

It is interesting to consider what the gambler can achieve if some limitation is put on the duration of play. It is assumed throughout this section that, for each f, the one-point measure $\delta(f)$ is in $\Gamma(f)$. The assumption could be dropped at the cost of minor complications, but the gain in generality would be only superficial.

Let $U_0 = u$; for each positive integer m let $U_m = \sup u(\pi)$ over policies $\pi = (\sigma, t)$ available in Γ at f, with $t \leq m$; let $U_\omega = \sup U_m$.

(A variant of this notation advantageous when indices larger than ω are to be used is this: $U'_1 = u$, $U'_\alpha = \sup u(\pi)$ over π in Γ at f, with the structure of $f_t < \alpha$. For $\alpha = m$, an integer, $U_m = U'_{m+1}$, and $U_\omega = U'_\omega$; so the two systems are essentially equivalent for $\alpha \leq \omega$. However, the definition of U_ω is ad hoc in the first system; extending the definition of U_m to general ordinal indices would lead to $U_\omega = U'_{\omega+1}$, and so U_ω would be left out, as would U'_α for every other limiting ordinal α.)

THEOREM 1. If $m < m'$, then $u = U_0 \leq U_m \leq U_{m'} \leq U_\omega \leq U$; and $\lim_m U_m = U_\omega$.

THEOREM 2. For γ ranging in $\Gamma(f)$, $U_{m+1}(f) = \sup \gamma U_m$.

In the concrete problems of later chapters, it is almost always true that $U_\omega = U$, but here is an example in which the equality fails.

> **Example 1.** Let: F be all pairs $f = (i, j)$ of nonnegative integers; γ be diffuse on the fortunes $(0, j)$; $\Gamma(0, 0) = \{\delta(0, 0), \gamma\}$ and, for other f, $\Gamma(f) = \{\delta(f), \delta(i + 1, j)\}$; $u = 0$ or 1 according as $i \leq j$ or $i > j$. It is evident that $U \equiv 1$, but that $U_m(0, 0) = U_\omega(0, 0) = 0$ for every m.

Just as in ordinary countably additive probability theory, if $v, v_1, v_2, \cdots$ are real-valued functions of fortunes and γ is a gamble such that, for every positive ϵ, $\gamma\{f\colon |v(f) - v_n(f)| > \epsilon\}$ converges to 0 as n approaches ∞, then v_n *converges to* v *in* γ*-probability*. The following is a finitely additive expression of the Lebesgue convergence theorems. (For another version, see (Dunford and Schwartz 1958, Theorem 7, p. 124).)

THEOREM 3. If uniformly bounded $v_1, v_2, \cdots$ converge to v in γ-probability, then $\lim \gamma v_n = \gamma v$. If the v_n are monotonic in n, then the converse is true, too.

THEOREM 4. If, for all f and γ in $\Gamma(f)$, U_m converges in γ-probability to U_ω, then $U_\omega = U$.

Proof: According to the basic Theorem 12.1, the following calculation, based on Theorems 2 and 3, provides the proof. For every f and γ in $\Gamma(f)$,

$$\gamma U_\omega = \lim_{m \to \infty} \gamma U_m \leq \lim_{m \to \infty} U_{m+1}(f) = U_\omega(f). \quad \blacklozenge \tag{1}$$

THEOREM 5. $U_\omega = U$ in each of the following seven cases:

(*a*) For every available γ there is a set of arbitrarily small γ-probability outside of which U_m converges uniformly to U_ω.

(*b*) U_m converges uniformly to U_ω.

(*c*) F is a compact topological space on which U_ω and U_m are continuous.

(*d*) F is a topological space on which U_ω and U_m are continuous, and, for each γ, γ of the complement of sufficiently large compact sets is arbitrarily small.

(*e*) F is a compact interval of real numbers, U_ω is continuous, and each U_m is monotonic.

(*f*) F is a not necessarily compact interval of real numbers, U_ω is continuous, each U_m is monotonic, and, for each γ, γ of the complement of sufficiently large compact subintervals of F is arbitrarily small.

(*g*) F has a sigma-field on which each γ is countably additive and with respect to which each U_m is measurable.

16. CONTINUOUS GAMBLING HOUSES

The purpose of this section is to illustrate that regularity conditions on a gambler's problem imply regularity conclusions about its solution. The hypotheses to be assumed in this section apply with trivial modifications to all the specific gambler's problems in this book, except to examples designed to exhibit irregularity. Later chapters will make no use of this section which is, in a sense, a continuation of Section 13. It can, therefore, freely be postponed or omitted.

Throughout this section:

(*a*) F is a compact metric space.

(*b*) The notion of a gamble γ is identified with its restriction to the Borel subsets $\mathcal{B}$ of F and also with the equivalence class of all gambles that agree with γ on $\mathcal{B}$.

(*c*) Every available γ is an element of Δ, the countably additive probabilities on $\mathcal{B}$.

(*d*) For each f, $\Gamma(f)$ is closed in the weak topology on Δ induced by the continuous functions on F, and $\Gamma(f)$ contains $\delta(f)$.

(*e*) The mapping $f \to \Gamma(f)$ is continuous for the Hausdorff-metric topology on 2^Δ, the nonvacuous, closed subsets of Δ, as in (Hausdorff 1957, Section 28). (The conclusions and proofs of this section remain valid under the weaker assumption that the mapping $f \to \Gamma(f)$ is upper semicontinuous in the sense of Kuratowski (1932). Namely, if f_n approaches f, $\gamma_n \in \Gamma(f_n)$, and γ_n approaches γ, then $\gamma \in \Gamma(f)$.)

(*f*) u is upper semicontinuous on F; that is, $\{f\colon z \leq u(f)\}$ is a closed subset of F for each real number z.

THEOREM 1. For every nonnegative integer m, U_m is upper semicontinuous and therefore Borel measurable; also $U = U_\omega$ and is Borel measurable. There is a sequence $\bar{\gamma}_0, \bar{\gamma}_1, \cdots$ of gamble-valued functions defined on F such that: $\bar{\gamma}_j$ is Borel measurable with respect to the Borel subsets of Δ; $\bar{\gamma}_j(f) \in \Gamma(f)$ for all f and j; $u(\pi) = U_{m+1}(f)$ if π is a policy for which $t \equiv m + 1$ and the first $m + 1$ gambles of σ are

$$\begin{aligned} \sigma_0 &= \bar{\gamma}_m(f), \\ \sigma_1(f_1) &= \bar{\gamma}_{m-1}(f_1), \\ &\cdots\cdots \\ \sigma_m(f_1, \cdots, f_m) &= \bar{\gamma}_0(f_m). \end{aligned} \tag{1}$$

Therefore Borel-measurable nearly optimal strategies exist.

The proof of the theorem will depend on seven lemmas.

LEMMA 1. If u is upper semicontinuous on F, then γu depends upper semicontinuously on the gamble γ.

Proof: Easy in consequence of the fact (Hausdorff 1957, Section 42) that u is the limit of a descending sequence of continuous functions. ◆

(Lemmas 2 through 5 and Lemma 7, in effect, apply to arbitrary compact-metric spaces F and Δ, and Lemma 6 applies to an arbitrary measurable space F and compact metric space Δ.)

Lemmas 2 through 7 refer to an upper-semicontinuous real-valued function v on Δ. Associated with v are: a real-valued function v^* with domain 2^Δ,

$$v^*(K) = \max_{\gamma \in K} v(\gamma); \tag{2}$$

and a mapping V of 2^Δ into itself,

$$V(K) = \{\gamma\colon \gamma \in K \text{ and } v(\gamma) = v^*(K)\}; \tag{3}$$

and a mapping $\tilde{V}$ of the Cartesian product of 2^Δ and the real numbers,

$$\tilde{V}(K, z) = \{\gamma\colon \gamma \in K \text{ and } v(\gamma) \geq z\}, \tag{4}$$

which, except when vacuous, is in 2^Δ.

LEMMA 2. v^* is upper semicontinuous and is therefore Borel measurable.

LEMMA 3. $\tilde{V}$ is upper semicontinuous (on the closed set where its value is nonvacuous) in the sense of Kuratowski (1932) and is therefore Borel measurable.

LEMMA 4. V is Borel measurable.

Proof: $V(K) = \tilde{V}(K, v^*(K))$. ◆

(We thank Donald Ornstein for a proof of Lemma 4 and Gerard Debreu for warning us away from a false short cut.)

LEMMA 5. The mapping $f \to V(\Gamma(f))$ is Borel measurable.

LEMMA 6. There is a Borel-measurable mapping $\bar{\gamma}$ of F into Δ such that $\bar{\gamma}(f) \in \Gamma(f)$ and $v(\bar{\gamma}(f)) = v^*(\Gamma(f))$.

Proof: Let w_j be a sequence of continuous, real-valued functions on Δ such that, if $\gamma \neq \gamma'$, there exists an i with $w_i(\gamma) \neq w_i(\gamma')$. Define W_j in analogy with V, and let $\Gamma_0(f) = V(\Gamma(f), \Gamma_{n+1}(f) = W_{n+1}(\Gamma_n(f))$. Each sequence of compact sets $\Gamma_n(f)$ decreases to a one-point set $\{\bar{\gamma}(f)\}$. Since $\bar{\gamma}(f) \in V(\Gamma(f))$, $v(\bar{\gamma}(f)) = v^*(\Gamma(f))$. Since $\{\bar{\gamma}(f)\} = \lim \Gamma_n(f)$, $\{\bar{\gamma}\}$ is a measurable mapping, and hence so is $\bar{\gamma}$. ◆

(Lemma 5 and, which is more important, Lemma 6 depend only on the measurability, not on the semicontinuity, of $\Gamma(f)$ in f. This semicontinuity is, however, essential to the next and final lemma).

LEMMA 7. $v^*(\Gamma(f))$ is upper semicontinuous in f.

Proof: v^* is not only upper semicontinuous but nondecreasing. ◆

Proof of Theorem 1: $U_0 = u$, which is upper semicontinuous by assumption. If v is the mapping $\gamma \to \gamma U_m$, then $U_{m+1}(f) = U_m^*(\Gamma(f))$. Therefore each U_m is upper semicontinuous, in view of Lemmas 1 and 7.

Each $\bar{\gamma}_m$ is obtained by applying Lemma 6.

According to part (*g*) of Theorem 15.5, $U_\omega = U$. Measurable, nearly optimal strategies are obtained by continuing (1) for large m with $\sigma_{m+r}(f_1, \cdots, f_{m+r}) = \delta(f_{m+r})$. ◆

If v is continuous, so is v^*. Therefore if u is continuous, so is U_m for each m. Are there then continuous nearly optimal strategies?

The hypothesis that $\delta(f) \in \Gamma(f)$ is almost superfluous, but, as the next chapter will make clear, if this hypothesis fails, Theorem 1 lacks interest and some deeper theorem should replace it.

It would be good to know to what extent the compactness of each $\Gamma(f)$ is essential to Theorem 1 and whether there are measurable nearly optimal strategies if Γ and u are assumed only to vary measurably with f.

3

STRATEGIES

1. INTRODUCTION.

This chapter, like the preceding one, is sufficiently general and abstract to apply to gambling problems based on any set of fortunes. Here the study of strategies, which in Chapter 2 played a role subsidiary to that of policies, is pursued. This chapter is entirely independent of Chapter 4, which might well be read first, and though this chapter is logically preliminary to much beyond Chapter 4, it is not quite so basic as most of Chapter 2.

2. THE UTILITY OF A STRATEGY

Strategies are closely akin to policies. The main reason for considering both is that, while policies are simpler to work with, strategies are conceptually simpler and optimal policies rarely exist.

If the kinship is to be close, a strategy, like a policy, must be assigned a utility. We do not intend to appraise a strategy σ in terms of the utilities of fortunes visited early, nor would it be appropriate to penalize σ for visits to fortunes of low utility however frequent, if these serve as stepping stones to fortunes of high utility. These considerations, together with the remark, in Section 2.11, that stop rules, as functions from H to the integers, are naturally partially ordered, suggest this definition:

$$u(\sigma) = \limsup_{t \to \infty} u(\sigma, t). \tag{1}$$

The "lim sup" here means, of course, the supremum of those numbers c such that for every t there is a t' with $t'(h) \geq t(h)$ for all h and $u(\sigma, t') \geq c$.

The definition (1) would be unsatisfactory if it did not permit this theorem:

THEOREM 1. For all σ,

$$u(\sigma) = \int u(\sigma[f])\, d\sigma_0(f). \tag{2}$$

Proof: Equation (2.10.2) says that

$$u(\sigma, t) = \int u(\sigma[f], t[f])\, d\sigma_0(f). \tag{3}$$

For a given stop rule t and positive ϵ, and for all f, let $t'(f)$ be a stop rule such that $t'(f) \geq t[f]$ and $u(\sigma[f], t'(f)) \geq u(\sigma[f]) - \epsilon$. There is a stop rule t' defined by the condition $t'(fh) = t'(f)(h) + 1$; equivalently, $t'[f] = t'(f)$ for all f. Clearly, $t' \geq t$, and, according to (3),

$$\begin{aligned} u(\sigma, t') &= \int u(\sigma[f], t'[f])\, d\sigma_0(f) \\ &= \int u(\sigma[f], t'(f))\, d\sigma_0(f) \\ &\geq \int u(\sigma[f])\, d\sigma_0(f) - \epsilon. \end{aligned} \tag{4}$$

Therefore, $u(\sigma) \geq \int u(\sigma[f])\, d\sigma_0(f)$.

On the other hand, for each ϵ and f there is a stop rule $t(f)$ such that $u(\sigma[f], t'(f)) \leq u(\sigma[f]) + \epsilon$ for $t'(f)$ as large as $t(f)$. If t is the stop rule for which $t(fh) = t(f)(h) + 1$ for all h, then $t[f] = t(f)$, and, for $t' \geq t$, $t'[f] \geq t(f)$ for all f. If $t' \geq t$,

$$\begin{aligned} u(\sigma, t') &= \int u(\sigma[f], t'[f])\, d\sigma_0(f) \\ &\leq \int u(\sigma[f])\, d\sigma_0(f) + \epsilon. \quad \blacklozenge \end{aligned} \tag{5}$$

3. STRATEGIC UTILITY

The notion of a gambler's ceasing to gamble after the partial history $p = (f_1, \cdots, f_n)$ can be given two technical interpretations. First, it can mean that he is employing a policy which terminates at time n after that partial history. Second, it can mean that the continuation of his strategy after p is such that, with probability 1, the fortune remains f_n time after time. As this section and the next will show, these two senses of ceasing are so close to each other in their effects that each policy can be imitated by a strategy. Moreover, as the next section makes evident, provided $\delta(f) \in \Gamma(f)$ for all f, a strategy σ that imitates π is available in Γ at f if π is.

A gambling house Γ is *leavable* if $\delta(f) \in \Gamma(f)$ for all f. We had so little spontaneous interest in nonleavable gambling houses that we were tempted to make leavability part of the official definition of a gambling house. We find, however, that the more general concept leads to better insight into the special one of main interest and, as Chapter 12 indicates, some problems of current interest are most naturally handled in the theory of not necessarily leavable houses.

In this chapter, and in some later contexts where the distinction makes a difference, we shall consider not only the utility U of a gambling house Γ but also the *strategic utility* V, where $V(f) = \sup u(\sigma)$ over all σ available in Γ at f. Comparison with the definition of U by (2.10.3) and with the definition of the utility of a strategy in Section 2 shows immediately that U majorizes V. Pictorially speaking, U reflects the possible achievements of a gambler who is free to go home when he pleases, whereas V represents the achievements of a gambler who is subject to detention in the gambling house for an arbitrarily long time and who must even select his σ before he is informed of the length of his detention. If the gambling house is leavable, then V and U are the same, as Corollary 2 states. When the gambling house is obviously leavable, we ordinarily write U rather than V. The notion of "gambler's problem" introduced in Section 2.10 is now modified to mean the determination of V and of optimal or nearly optimal strategies.

With the aid of Theorem 2.1, it is as easy to prove that V is excessive for Γ as it was to prove the corresponding fact (Theorem 2.14.1) for U. Similarly, if Q is excessive for Γ and $Q(\sigma) \geq u(\sigma)$ for all σ available in Γ, then it is easy to paraphrase the proof of Theorem

2.12.1 to conclude that Q majorizes V. These remarks and one lemma easily pave the way to extensions to V of the basic facts about U as a function excessive for Γ.

For each nonvacuous partial history $p = (f_1, \cdots, f_n)$, let $l(p) = f_n$.

LEMMA 1. For all f and all σ available in Γ at f, $V(\sigma) \geq u(\sigma)$.

Proof: The two general formulas,

$$u(\sigma) = \int u(\sigma[p])\, d\eta(\pi)(p) \tag{1}$$

and

$$\int V(l(p))\, d\eta(\pi)(p) = \int V(g)\, d\gamma(\pi)(g) = V(\pi), \tag{2}$$

are applicable to any bounded functions u and V of fortunes, as is easily verifiable by induction. (Compare with the related formula (2.9.6) and Theorem 2.1.) In the special case that V is the V of a Γ, and σ is in Γ at some f, $V(l(p)) \geq u(\sigma[p])$ for all p. Consequently, $V(\sigma, t) = V(\pi) \geq u(\sigma)$ for all t; so $V(\sigma) \geq u(\sigma)$. ◆

THEOREM 1. V is the smallest of all those functions Q that are excessive for Γ and for which $Q(\sigma) \geq u(\sigma)$ for every σ available in Γ (at any f).

COROLLARY 1. If $V \geq u$, then V is U.

COROLLARY 2. If Γ is leavable, V is U.

Let Γ^L, the *leavable closure* of Γ, be the smallest leavable Γ that includes Γ. That is, $\Gamma^L(f) = \Gamma(f) \cup \{\delta(f)\}$ for every fortune f.

COROLLARY 3. The U of any gambling house Γ is the V of Γ^L.

As Theorem 1 and its corollaries suggest, it would have been mathematically more efficient to begin with the concept V and regard that of U as auxiliary. We thought it better, however, to present U

first—in imitation of our own experience—because, in the problems that interest us most, Γ is leavable.

COROLLARY 4. If σ is a strategy available in Γ at f, and t and t' are stop rules with $t \leq t'$, then

$$V(f) \geq \sigma_0 V \geq V(\sigma, t) \geq V(\sigma, t') \geq V(\sigma) \geq u(\sigma). \tag{1}$$

4. STAGNANT STRATEGIES

This section shows that, for every fortune f', there is a strategy σ' that imitates the act of accepting the initial fortune f' without actually embarking on a policy and, for every policy π, there is some strategy σ that imitates π. For every u, $u(\sigma') = u(f')$ and $u(\sigma) = u(\pi)$. The strategy σ' is available in every leavable Γ at f', and, for every leavable Γ and fortune f, σ is available in Γ at f if π is. These facts go beyond the assertion of Corollary 3.2 that U is V for every leavable Γ.

If, for some h and σ, every finitary set has probability 1 or 0 according as h is or is not in the set, σ (or, more accurately, the measure determined by σ) is the *one-point measure* $\delta(h)$ at h. If F has more than one element, no one-point set of histories is finitary according to Corollary 2.7.1. Yet the finitary sets (and even those of finite structure) are adequate to separate points. Therefore, if $\delta(h) = \delta(h')$, then $h = h'$. (Of course, different strategies σ can determine the same measure σ, which can, in particular, be a one-point measure.)

THEOREM 1. The following conditions are equivalent:

(*a*) $\sigma = \delta(h)$.

(*b*) $\sigma g = g(h)$ for all inductively integrable g.

(*c*) $\sigma g = g(h)$ for all inductively integrable g that depend on only finitely many coordinates, that is, are of finite structure.

(*d*) $\sigma_0 = \delta(f_1)$ and $\sigma_n(f_1, \cdots, f_n) = \delta(f_{n+1})$ for all positive n (where $h = (f_1, f_2, \cdots)$).

(*e*) For each n, the probability under σ that the nth fortune is f_n is 1.

(*f*) For each stop rule t, the probability under σ that $t = t(h)$ is 1.

(*g*) For each t, the σ-probability that $f_t = f_{t(h)}$ is 1.

(*h*) $\sigma_0 = \delta(f_1)$ and $\sigma[f_1] = \delta(h')$, where $h' = (f_2, f_3, \cdots)$.

Of course, the components of the strategy σ along histories other than h have no bearing on whether the measure σ is $\delta(h)$.

Let h_f be the history $(f, f, f, \cdots)$. If the measure σ is $\delta(h_f)$, the strategy σ *stagnates immediately at* f. For σ to stagnate immediately at f it is necessary and sufficient that $u(\sigma, t) = u(f)$ for all t and u. If the conditional strategy $\sigma[f_1, \cdots, f_n]$ (defined in Section 2.5) is $\delta(h_{f_n})$, that is, if $\sigma[f_1, \cdots, f_n]$ stagnates immediately at f_n, the strategy σ *stagnates (at n) at* $(f_1, \cdots, f_n)$. The gambler here decides after $(f_1, \cdots, f_n)$ to remain at f_n with probability 1 on his $(n + 1)$st gamble, to do so again at his $(n + 2)$nd if he finds himself at f_n (which he is almost certain to), and so on.

Stagnation is closely akin to termination. In particular, if stagnation occurs at a given partial history $(f_1, \cdots, f_n)$ and if the stop rule t stops then or later, the terminal fortune given the first n coordinates $(f_1, \cdots, f_n)$ is f_n with probability 1, as the next theorem formally asserts.

THEOREM 2. If σ stagnates at $(f_1, \cdots, f_n)$ and if $tf_1f_2 \cdots f_n \geq n$, then

$$\sigma[f_1, \cdots, f_n]\{h: f_tf_1 \cdots f_n(h) = f_n\} = 1. \tag{1}$$

Proof: The measure concerned is the one-point measure on the history $h = (f_n, f_n, \cdots)$, and the set concerned contains this h, since $f_tf_1 \cdots f_n(h) = f_n$. ◆

With each σ associate the (possibly) incomplete stop rule t_σ such that $t_\sigma(h)$ is the first n, if any, for which σ stagnates at $(f_1, \cdots, f_n)$; failing such an n, $t_\sigma(h) = \infty$. We intend that $t_\sigma(h) = 1$ if σ stagnates immediately or at time 1 at f_1. According to the definition of t_σ, $t_{\sigma[f]} = t_\sigma f - 1$, unless $\sigma[f]$ stagnates immediately at f, in which case $t_{\sigma[f]} = t_\sigma f = 1$.

As the following theorem says formally, it is immaterial when a gambler leaves the gambling house once he has decided to remain inactive.

THEOREM 3. For any utility u and policy (σ, t),

$$u(\sigma, t) = u(\sigma, \min(t_\sigma, t)).$$

Proof: Recall the convention that $u(\sigma[f], 0) = u(f)$ implied in the paragraph containing (2.10.2), so that (2.3) is valid without exception. Then proceed by induction on the structure $\alpha(f_t)$, calculating thus.

$$
\begin{aligned}
(2) \qquad u(\sigma, t) &= \int u(\sigma[f], t[f]) \, d\sigma_0(f) \\
&= \int u(\sigma[f], \min(t_{\sigma[f]}, t[f])) \, d\sigma_0(f) \\
&= \int u(\sigma[f], \min(t_\sigma f - 1, tf - 1)) \, d\sigma_0(f) \\
&= \int u(\sigma[f], \min(t_\sigma f, tf) - 1) \, d\sigma_0(f) \\
&= \int u(\sigma[f], \min(t_\sigma, t)f - 1) \, d\sigma_0(f) \\
&= \int u(\sigma[f], (\min(t_\sigma, t))[f]) \, d\sigma_0(f) \\
&= u(\sigma, \min(t_\sigma, t)). \quad \blacklozenge
\end{aligned}
$$

If t_σ is a (complete) stop rule, σ is *stagnant.*

COROLLARY 1. If there is a stop rule t for which $t \geq t_\sigma$, then σ is stagnant and $u(\sigma, t) = u(\sigma, t_\sigma)$.

COROLLARY 2. If σ is stagnant, then $u(\sigma) = u(\sigma, t_\sigma)$.

Just as it is immaterial to $u(\pi)$ where termination occurs, provided it is not before stagnation, so, too, the behavior of the strategy after termination is immaterial.

THEOREM 4. If $\sigma_0' = \sigma_0$ and if, for all $h = (f_1, f_2, \cdots)$ and all $n < t(h)$, $\sigma_n(f_1, \cdots, f_n) = \sigma_n'(f_1, \cdots, f_n)$, then $u(\sigma', t) = u(\sigma, t)$ for all u.

For each history $h = (f_1, f_2, \cdots)$ and stop rule t, if $t(h) > n$, *t terminates after* $p = (f_1, \cdots, f_n)$; if $t(h) \leq n$, *t terminates by* p.

For each policy $\pi = (\sigma, t)$, there is a strategy σ' closely equivalent to π in its effects. The natural candidate for σ' is σ_π defined by $\sigma_\pi(p) = \sigma(p)$ or $\delta(l(p))$ according as t terminates after p or by p. (Recall that $l(p)$ is the last coordinate of p.)

THEOREM 5. For all π, σ_π is a stagnant strategy with $t_{(\sigma_\pi)} \leq t$; σ_π is available in any leavable gambling house Γ at f for which σ is; and $u(\sigma_\pi, t) = u(\pi) = u(\sigma, t) = u(\sigma_\pi)$.

If the set of histories where t_σ is finite has (Eudoxian) probability 1 under σ, then σ *stagnates almost certainly;* $u(\sigma, t_\sigma)$ is then defined in accordance with Section 2.11.

THEOREM 6. If σ stagnates almost certainly, $u(\sigma) = u(\sigma, t_\sigma)$. In fact, $\lim_t u(\sigma, t) = u(\sigma, t_\sigma)$.

Proof: If $t \geq t_\sigma$ on a finitary set A with complement $\bar{A}$ and $\sigma\bar{A} \leq \epsilon$, let $t' = \min(t, t_\sigma)$.

$$
\begin{aligned}
|u(\sigma, t) - u(\sigma, t_\sigma)| &= |u(\sigma, t') - u(\sigma, t_\sigma)| \\
&= |\sigma[u(f_{t'}) - u(f_{t_\sigma})]| \\
&\leq (\sigma\bar{A})[\sup u - \inf u].
\end{aligned}
\tag{3}
$$

The right side of (3) is arbitrarily small for all sufficiently large t. ♦

THEOREM 7. If u assumes only 0 and 1 as values, and if σ stagnates at $(f_1, \cdots, f_n)$ whenever $u(f_n) = 1$, then $u(\sigma, t)$ is nondecreasing in t; consequently, $u(\sigma) = \lim_t u(\sigma, t)$.

5. OPTIMAL STRATEGIES

This section and the following three form a sort of subchapter about optimal strategies. This section introduces and makes a preliminary study of optimal strategies. Optimality is shown to be the intersection of two other properties. The next two sections study the two properties separately. A fourth section brings together some of the conclusions about optimal strategies implicit in the first three.

From the point of view that led to the definition of U in Section 2.10, a good behavior on the part of a gambler with initial fortune f is to adopt a policy π for which $u(\pi)$ is very close to $U(f)$ or to settle for $u(f)$, if that is very close to $U(f)$. If, however, all behavior is to be expressed in terms of strategies, which the preceding sections indicate can be done without mathematical loss, then good behavior at f is any strategy σ available in Γ at f for which $u(\sigma)$ is very close to $V(f)$.

So far as optimal, as opposed to merely good, behavior is concerned, we here focus attention on optimal strategies only. Optimal policies are rare and have not proved very interesting. In studying optimal strategies, we emphasize the general theory of not necessarily leavable gambling houses, so that V is not necessarily the same as U. At the same time, examples and counterexamples will be chosen from among leavable gambling houses.

A strategy σ available in Γ at f is *optimal* at f if $u(\sigma) = V(f)$. Though this does seem the almost inevitable definition of "optimal", an optimal strategy is not necessarily quite without blemish. Even in the presence of countable additivity, an optimal strategy might permit bad behavior with probability 0; in the absence of countable additivity, bad behavior may occur along every partial history.

Example 1. Let F consist of the positive integers i. Let each $\Gamma(i)$ be the same for all i and consist of every gamble on F. Let $u(i) = 1 - i^{-1}$. A strategy σ that begins with a diffuse gamble γ and then stagnates is available at every fortune i and is optimal there. Yet $\sigma[j]$, the conditional continuation of σ after the initial gamble γ results in j, is not optimal.

A strategy σ that is optimal at f might be called thoroughly optimal at f if, for each partial history p, the conditional strategy $\sigma[p]$ is optimal at $l(p)$. Some concepts intermediate between optimal and thoroughly optimal are suggested by the expression "almost thoroughly optimal", but we refrain from exploration of any concepts stronger than that of optimality. The strategy σ in Example 1 is of course optimal but not thoroughly optimal, though the gambling house Γ of that example abounds in thoroughly optimal strategies.

The following reformulation of optimality is obvious in the light of Corollary 3.4.

THEOREM 1. A σ available in Γ at f is optimal there if and only if $V(f) = V(\sigma)$ and $V(\sigma) = u(\sigma)$.

Theorem 1 displays the strategies optimal for u in Γ at f among those available in Γ at f as just those that have two properties that can be viewed as aspects of optimality. The next section investigates the strategies which have one of these properties, and the one after that, those which have the other.

The next theorem is immediate from Theorem 3.1.

THEOREM 2. If $u \leq u' \leq V$, then $V' = V$, and a strategy optimal for u in Γ at f is also optimal for u' in Γ at f.

A strategy can, of course, be optimal for u' without being optimal for u.

6. THRIFTY STRATEGIES

The purpose of this section is to continue the exploration of optimal strategies by studying those strategies σ that are *thrifty* for the V of a gambling house at some fortune f, that is, those for which $V(f) = V(\sigma)$.

It is convenient, however, to replace this by a superficially more general problem: Among the strategies σ for which an arbitrary bounded function V is excessive, which σ are thrifty for V at f? Therefore, whenever a σ, V, and f are mentioned in the rest of this section, it is to be understood that V is excessive for σ at f, and it is also to be remembered that the conclusions have an immediate application to the situation in which V is the V of a gambling house Γ and σ is available in Γ at f. Examples and counterexamples will be so constructed that the strategy σ will be available in a leavable gambling house Γ the U of which is V.

All but the final inequality in Corollary 3.4 depend only on the fact that V is excessive for Γ at f. Therefore, for $t \leq t'$,

$$V(f) \geq \sigma_0 V \geq V(\sigma, t) \geq V(\sigma, t') \geq V(\sigma). \tag{1}$$

A reformulation of thriftiness flows immediately from (1).

LEMMA 1. $V(\sigma) = V(f)$ if and only if $V(\sigma, t) = V(f)$ for all stop rules t.

If $\gamma V \geq V(f)$, γ *conserves* V *at* f; if σ_0 conserves V at f and $\sigma_j(f_1, \cdots, f_j)$ conserves V at f_j for $1 \leq j < n$, σ *conserves* V *at* f *along* $h = (f_1, f_2, \cdots)$ *up to* n. For each stop rule t, the event that σ conserves V at f up to t is finitary. If, for every t, this event has σ-probability 1, then σ *persistently conserves* V *at* f.

LEMMA 2. If the probability that σ conserves V at f up to t is $1 - \epsilon$, then $V(\sigma, t) \geq V(f) - \epsilon(\sup V - \inf V)$.

Proof: Consider ϵ as a function of σ and f and use induction on the structure of f_t. ◆

(We thank Manish Bhattacharjee for Lemma 2.)

THEOREM 1. If σ persistently conserves V at f, then $V(\sigma) = V(f)$; that is, σ is thrifty for V at f.

Proof: Apply Lemmas 1 and 2. ◆

(If σ persistently conserves V at f, then the set of histories along which σ conserves V up to ∞, that is, up to every n, has outer probability 1. But its inner probability may be 0 even if F is finite; even more, it may include no nonempty finitary subset.

Example 1. Let $F = \{0, 1, 2\}$; $\Gamma(0) = \{\delta(0)\}$, $\Gamma(1) = \{\delta(0), \delta(1), \delta(2)\}$, $\Gamma(2) = \{\delta(2)\}$; $u(0) = u(1) = 0$, $u(2) = 1$; $\sigma_0 = \delta(2)$, $\sigma_n(f_1, \cdots, f_n) = \delta(f_n)$ if $f_n = 0$ or 2, and $\delta(0)$ if $f_n = 1$.)

A converse to Theorem 1 for countably additive situations is indicated below, but the full converse is in conflict with the next example.

Example 2. Let F consist of the ordered pairs of integers $f = (i, j)$, with $i = 1$ and 2 and $j \geq 1$. Let $\Gamma(1, 1)$ consist of $\delta(1, 1)$ and a diffuse γ on the fortunes $(1, j)$ (that is, for all k,

$$\gamma\{f\colon i = 1 \text{ and } j > k\} = 1);$$

for all other f whose first coordinate is 1, let $\Gamma(f)$ consist of $\delta(f)$ and $\delta(g)$ for all g whose first coordinate is 2; for all f whose first coordinate is 2, let $\Gamma(f)$ contain only $\delta(f)$. Let $u(i, j) = 0$ or $1 - j^{-1}$ according as $i = 1$ or 2. Then $U(i, j) = 1$ for $i = 1$, and $1 - j^{-1}$ for $i = 2$; γ is the only nontrivial gamble that conserves U. Any σ in Γ that first uses γ and then, when the first coordinate is 1, uses any nontrivial gamble that does not decrease the second coordinate is optimal at (1, 1) and therefore thrifty at (1, 1), but σ does not persistently conserve U.

Though σ need not persistently conserve in order to be thrifty, it must approximately do so. A gamble γ *ϵ-conserves V at f* if $\gamma V \geq V(f) - \epsilon$; if σ_0 ϵ-conserves V at f and $\sigma_j(f_1, \cdots, f_j)$ ϵ-conserves V at f_j for $1 \leq j < n$, then σ *ϵ-conserves V at f along $h = (f_1, f_2, \cdots)$ up to n*. If, for each stop rule t, σ ϵ-conserves V at f up to t with σ-probability 1, then σ *persistently ϵ-conserves V at f*.

LEMMA 3. If α is the σ-probability that σ ϵ-conserves V at f up to t, then

$$V(\sigma, t) \leq V(f) - (1 - \alpha)\epsilon. \tag{2}$$

Proof: Use induction on the structure of f_t thus. If σ_0 does not ϵ-conserve V at f, then $V(\sigma, t) \leq \sigma_0 V$ according to (1), and $\sigma_0 V < V(f) - \epsilon$. If σ_0 is ϵ-conserving, use formula (2.10.2), Theorem 2.9.3, and the inductive hypothesis. ◆

THEOREM 2. If σ is thrifty for V at f, then, for each positive ϵ, σ is persistently ϵ-conserving. (Of course, the assertion cannot be made for $\epsilon = 0$.)

Proof: Apply Lemmas 1 and 3. ◆

COROLLARY 1. If σ is thrifty for V at f and the partial history $(f_1, \cdots, f_n)$ occurs with positive probability, then the gamble $\sigma_n(f_1, \cdots, f_n)$ conserves V at f_n.

Theorem 1 states a sufficient condition that σ be thrifty, which was seen in Example 2 not to be necessary. Theorem 2 states a necessary condition, which, as the next example shows, is not sufficient. We have not succeeded in closing this gap in the theory of thriftiness and optimal strategies.

Example 3. Let F be the set of ordered pairs of positive integers $f = (i, j)$, with $i \geq j$. Define a $\gamma(f)$ for each f thus: $\gamma(1, 1)$ is any diffuse gamble on the fortunes $(i, 1)$; $\gamma(i, j) = \delta(i, j + 1)$ if $j < i$; $\gamma(i, i) = \delta(i, i)$ if $i > 1$. Let $\Gamma(f) = \{\delta(f), \gamma(f)\}$ and $u(f) = -j/i$. It is easy to verify that $U(1, 1) = 0$ and $U(f) = u(f)$ if $f \neq (1, 1)$. Define a stagnant σ available in Γ at $(1, 1)$ thus:

$\sigma_0 = \gamma(1, 1)$, $\sigma_n(f_1, \cdots, f_n) = \gamma(f_n)$. Define $t(h)$ as the first coordinate of f_1. Though σ is persistently ϵ-conserving, $U(\sigma) \leq U(\pi) = -1 < U(1, 1) = 0$.

In the presence of countable additivity, σ persistently ϵ-conserves V at f for every positive ϵ if and only if it does so for $\epsilon = 0$. In this important environment, then, Theorem 1 and Theorem 2 establish a condition necessary and sufficient for thriftiness. In order to avoid a technical discussion of measurability, the presentation of this fact is here limited to a special case, which is more than adequate for the applications in Chapters 5 and 6.

As usual, a gamble γ is *discrete* if there is a countable sequence of fortunes $f_1, f_2, \cdots$ for which $\Sigma\, \gamma\{f_i\} = 1$. A strategy σ is *(persistently) discrete* if, for every stop rule t, with σ-probability 1, the only gambles $\sigma(p)$ used prior to termination are discrete.

THEOREM 3. If σ is discrete, then $V(\sigma) = V(f)$ (that is, σ is thrifty for V at f) if and only if σ_0 conserves V at f and $\sigma(p)$ conserves V at $l(p)$ for every nonvacuous partial history p that occurs with positive σ-probability.

7. EQUALIZING STRATEGIES

This section continues the study of optimal strategies for a gambling house Γ and in Theorem 2 characterizes those σ that *equalize* u and V, that is, those for which $u(\sigma) = V(\sigma)$.

Consonant with Section 2.6, for each function g on H and each $p = (f_1, \cdots, f_n)$, gp evaluated at $h' = (f'_1, f'_2, \cdots)$ is

$$g(f_1, \cdots, f_n, f'_1, f'_2, \cdots).$$

If g is inductively integrable, B is a subset of partial histories, and $A = \{h\colon p_t(h) \in B\}$, then

$$\int_A g\, d\sigma = \int_A \sigma[p_t(h)](gp_t(h))\, d\sigma(h) \tag{1}$$
$$= \int_B \sigma[p](gp)\, d\eta(p),$$

where η is the distribution of p_t under σ, as is verifiable by induction on the structure of p_t.

LEMMA 1. If w is the indicator function of a set of fortunes and σ is any strategy (in Γ or not) for which $w(\sigma) > 0$, then, for every $\delta > 0$, there is a partial history p for which $w(\sigma[p]) > 1 - \delta$.

Proof: For any positive ϵ and ϵ', and any stop rule t'', there is a $t' \geq t''$ for which

$$(2) \qquad \sigma\{h\colon w(\sigma[p_{t'}(h)]) \geq \epsilon'\} < w(\sigma) + \epsilon,$$

as the next two paragraphs show.

Choose $t' \geq t''$ such that $w(\sigma) + \frac{1}{2}\epsilon \geq w(\sigma, t') \geq w(\sigma) - \frac{1}{4}\epsilon\epsilon'$. Let $A = \{h\colon w(f_{t'}(h)) = 0$ and $w(\sigma[p_{t'}(h)]) \geq \epsilon'\}$, and let $\bar{A}$ be its complement.

If t' does not satisfy (2), then $\sigma\{h\colon w(\sigma[p_{t'}(h)]) \geq \epsilon'\} \geq w(\sigma) + \epsilon \geq w(\sigma, t') + \frac{1}{2}\epsilon = \sigma\{h\colon w(f_{t'}(h)) \neq 0\} + \frac{1}{2}\epsilon$; so $\sigma A \geq \frac{1}{2}\epsilon$. Clearly,

$$w(\sigma, t') = \int_{\bar{A}} w(f_{t'}(h))\, d\sigma(h).$$

For each p, let $t^*(p)$ be a stop rule for which $w(\sigma[p], t^*(p)) \geq \frac{2}{3}w(\sigma[p])$. Let t^{**} be the composition of t' with the family t^* as in (2.9.2). According to (1),

$$\begin{aligned}
\int_A w(f_{t^{**}})\, d\sigma &= \int_A \sigma[p_{t'}(h)]w(f_{t^*(p_{t'}(h))})\, d\sigma(h) \\
&= \int_A w(\sigma[p_{t'}(h)], t^*(p_{t'}(h)))\, d\sigma(h) \\
&\geq \tfrac{2}{3}\epsilon'\sigma A \\
&\geq \tfrac{1}{3}\epsilon\epsilon'.
\end{aligned}$$

If h is in A, let $t(h) = t^{**}(h)$; if h is not in A, let $t(h) = t'(h)$. This defines a stop rule t, and

$$\begin{aligned}
w(\sigma, t) &= \int_A w(f_t)\, d\sigma + \int_{\bar{A}} w(f_t)\, d\sigma \\
&= \int_A w(f_{t^{**}})\, d\sigma + w(\sigma, t') \\
&\geq \tfrac{1}{3}\epsilon\epsilon' + w(\sigma) - \tfrac{1}{4}\epsilon\epsilon' \\
&> w(\sigma),
\end{aligned}$$

which is impossible. Therefore t' does satisfy (2).

For any stop rule t',

$$w(\sigma) = \int w(\sigma[p_{t'}(h)])\, d\sigma(h), \tag{3}$$

in view of (3.1).

In consequence of (2) and (3), were $w(\sigma[p])$ bounded by $1 - \delta$, $w(\sigma)$ would not exceed $\epsilon'(1 - w(\sigma) - \epsilon) + (1 - \delta)(w(\sigma) + \epsilon) < w(\sigma)$ for ϵ' and ϵ sufficiently close to 0. ◆

(Incidentally, Lemma 1 can, with some effort, be strengthened into a general (that is, finitely additive) version of Lévy's (1937, Section 41) martingale theorem, and even Doob's (1953, Theorem VII 4.1) martingale convergence theorem can be so generalized.)

For each $\epsilon > 0$, a fortune f is *ϵ-unstable* if $u(f) > V(f) + \epsilon$. Under any available strategy, the gambler's fortune is unlikely to be ϵ-unstable far in the future.

THEOREM 1. Let w_ϵ be the indicator of the ϵ-unstable fortunes. If σ is available in Γ, then $w_\epsilon(\sigma) = 0$ for each positive ϵ.

Proof: Otherwise, according to Lemma 1, for each $\delta > 0$ there is a strategy σ', available in Γ, for which $w_\epsilon(\sigma') > 1 - \delta$. So, for each stop rule t, there is a stop rule $t' \geq t$ such that, with σ'-probability at least $1 - \delta$, $u(f_{t'}) > V(f_{t'}) + \epsilon$. Consequently, $u(\sigma', t') \geq V(\sigma', t') + \epsilon - (M - m + \epsilon)\delta \geq V(\sigma') + \epsilon - (M - m + \epsilon)\delta$, where M and m are upper and lower bounds for u and hence for V. For δ sufficiently small, $u(\sigma') \geq V(\sigma') + \frac{1}{2}\epsilon$, contradictory to Lemma 3.1. ◆

A fortune f is *ϵ-adequate* if $u(f) \geq V(f) - \epsilon$ and *adequate* if it is 0-adequate, or $u(f) \geq V(f)$.

THEOREM 2. Let v_ϵ be the indicator of the ϵ-adequate fortunes. For σ available in Γ, $u(\sigma) = V(\sigma)$, or σ equalizes u and V, if and only if for every $\epsilon > 0$, $v_\epsilon(\sigma) = 1$.

Proof: As is trivial to verify, this condition implies $u(\sigma) \geq V(\sigma)$. Lemma 3.1 asserts the reverse inequality. On the other hand, if $u(\sigma) = V(\sigma)$ and $\epsilon > 0$, Theorem 1 plainly implies that $v_\epsilon(\sigma) = 1$. ◆

COROLLARY 1. If σ is available in Γ and almost certainly stagnates, then $u(\sigma) = V(\sigma)$ if and only if, for each $\epsilon > 0$, the fortune at stagnation is ϵ-adequate with probability 1.

The "ϵ" in Theorem 2, unlike the one in Theorem 6.2, enters not because of finite additivity.

Example 1. Let F be the positive integers; $\Gamma(f) = \{\delta(f), \delta(f+1)\}$; and $u(f) = 1 - f^{-1}$. Then $U \equiv 1$; no fortune is 0-adequate; the strategy σ of moving forward step by step from any f_0 is optimal, and $u(\sigma) = U(\sigma) = 1$.

But if F is finite, the "ϵ" in Theorem 2 can be eliminated.

COROLLARY 2. Suppose that F is finite. Then $u(\sigma) = V(\sigma)$ if and only if, for every stop rule t and positive δ, there is a stop rule $t' \geq t$ such that, with σ-probability at least $1 - \delta$, $f_{t'}$ is adequate.

In a countably additive application, such as Corollary 2, the t in Theorem 2 need range over bounded stop rules only, but this is not so in general.

Example 2. Let: F be the pairs of integers $f = (i, j)$, with $i = 0, 1$ and $j \geq 0$; γ be a diffuse gamble on the fortunes $(1, j)$; $\Gamma(f) = \{\delta(0, j), \delta(1, j), \gamma\}$ for all f; $\sigma_0 = \gamma$, and $\sigma_n(f_1, \cdots, f_n) = \delta(f_n)$ or $\delta(0, j_n)$ according as $n < j_1$ or $n \geq j_1$; and $u(i, j) = i$. Check that: σ is stagnant and in Γ at every f; $u(\sigma) = 0$; $U \equiv 1$; and yet $\sigma\{h\colon f_n \text{ is adequate}\} = 1$ for all $n \geq 1$.

8. OPTIMAL STRATEGIES AGAIN

As is emphasized by Theorem 5.1, every fact about thrifty and equalizing strategies has an implication for optimal strategies. A few of the implications of the preceding three sections that are not quite automatic, and some other facts, are here put on record.

THEOREM 1. If no gamble in $\Gamma(f)$, other than possibly $\delta(f)$, conserves V at f, and if $u(f) < V(f)$, then there is no optimal strategy for u in Γ at f.

THEOREM 2. If σ is available in Γ at f, persistently conserves V at f, and almost surely stagnates at an adequate fortune, then σ is optimal at f.

Proof: Apply Corollary 7.1, Theorem 6.1, and Theorem 5.1. ◆

There may be no optimal strategies even if each $\Gamma(f)$ includes only a finite number of γ and each γ is countably additive.

Example 1. Let F be all integers greater than -2. Let $u(0) = 1$, and $u(i) = 0$ for $i \neq 0$. For $i = 0$ and $i = -1$, $\Gamma(i)$ contains only $\delta(i)$. For all other i, $\Gamma(i) = \{\delta(i), \delta(i + 1), \gamma(i)\}$, where $\gamma(i)$ assigns to 0 probability $1 - i^{-1}$ and to -1 probability i^{-1}. Then $U(i) = 1$ for $i \geq 0$, but for $i \geq 1$, clearly no optimal strategies exist.

Even if F is finite, provided that it has as many as three elements, there may exist no optimal strategies at some f.

Example 2. Let $F = \{0, 1, 2\}$; $\Gamma(0) = \{\delta(0)\}$, $\Gamma(1) = \{\delta(1), (1 - n^{-1})\delta(2) + n^{-1}\delta(0)$ for $n = 1, 2, \cdots\}$, $\Gamma(2) = \{\delta(2)\}$; $u(i) = i$, $i = 0, 1, 2$. There is no optimal strategy available in Γ at 1.

The notation $|p|$ is sometimes convenient for the number of coordinates of the partial history p, as in the proof of the next theorem.

THEOREM 3. For each f, there is a strategy σ available in Γ at f such that, for every partial history p (including the null one), $u(\sigma[p]) \geq \inf U$, and therefore $\inf V = \inf U$.

Proof: Without real loss in generality, assume that the infimum of U is 0. For each f and n, there is then a policy $\pi_n(f)$ for which $u(\pi_n(f)) > -1/n$. Construct a new sequence of families of policies $\pi_n^*(f) = (\sigma_n^*(f), t_n^*(f))$ by induction, thus: Let $\pi_1^*(f)$ be $\pi_1(f)$, and for each f and $n \geq 1$, let $\pi_{n+1}^*(f)$ be the policy obtained by following the policy $\pi_n^*(f)$ until it terminates, say at f', and by then continuing with the family of policies $\pi_{n+1}(f')$ in accordance with Section 2.9. For each f, the sequence of policies $\pi_1^*(f)$, $\pi_2^*(f)$, $\cdots$ is such that for all n, all $m > n$, and all partial histories p, if $t_n^*(f)$ does not terminate by p, then $\sigma_n^*(f)(p) = \sigma_m^*(f)(p)$. In particular, for each p and all sufficiently large n, all $\sigma_n^*(f)$ prescribe the same gamble after p, say $\sigma(f)(p)$.

To verify that $u(\sigma(f)[p]) \geq 0$, let t' be an arbitrary stop rule and n a large integer. Let $m(h)$ be the first integer as large as n for which $t_{m(h)}^*(f)(ph) \geq t'(h) + |p|$, and let $t(h) = t_{m(h)+1}^*(f)(ph) - |p|$. Plainly, t is a stop rule, and $t \geq t'$. As is easy to see, $u(\sigma(f)[p], t) \geq -(n + 1)^{-1}$. For $(\sigma(f)[p], t)$ is the composition of a policy π'' with the family $\pi_{m(h)+1}(f')$, and $u(\pi_{m(h)+1}(f')) \geq -(n + 1)^{-1}$. In fact, $\pi'' = (\sigma(f)[p], t_{m(h)}^*(f)(ph))$. ◆

COROLLARY 1. If V is constant, there is, at each f, an optimal strategy available in Γ.

THEOREM 4. There is, at each f, a σ available in Γ that equalizes u and V, that is, for which $u(\sigma) = V(\sigma)$.

Proof: Let $u' = u - V$. Then $V' \equiv 0$, and Corollaries 1 and 3.4 apply. ◆

LEMMA 1. If u is an indicator, σ is a strategy available at some f, and $u(\sigma[p]) \geq \delta$ for some $\delta > 0$ and all partial histories p, then $u(\sigma) = 1$.

Proof: For any $\epsilon > 0$, there is a stop rule t such that $u(\sigma) - \epsilon \leq u(\sigma, t)$ and for all $t' \geq t$, $u(\sigma, t') \leq u(\sigma) + \epsilon$. Let $t' = t$ where $u(f_t) = 1$. Elsewhere let t' be the composition of t with a family $t^*(\cdot)$, where $u(\sigma[p], t^*(p)) > \delta - \epsilon$. Then $u(\sigma) + \epsilon \geq u(\sigma, t') \geq u(\sigma, t) + [1 - u(\sigma, t)](\delta - \epsilon) \geq [u(\sigma) - \epsilon] + [1 - u(\sigma) - \epsilon](\delta - \epsilon)$. So $u(\sigma) \geq u(\sigma) + [1 - u(\sigma)]\delta + 0(\epsilon)$, which is impossible unless $u(\sigma)$ is 1. ◆

THEOREM 5. If u is an indicator function and V is bounded away from 0, then $V \equiv 1$, and at each f there is an optimal strategy.

Proof: Apply Theorem 3 and Lemma 1. ◆

9. STATIONARY FAMILIES

A family of strategies $\bar{\sigma}$ was defined as a mapping from fortunes to strategies; $\bar{\sigma}$ is *stationary* if, for some gamble-valued function $\bar{\gamma}$, $\bar{\sigma}(f)(p)$ is $\bar{\gamma}(l(p))$ unless p is the empty partial history, in which case it is $\bar{\gamma}(f)$. Intuitively, it would seem that a gambler at his nth decision ought not to benefit from knowledge of his past fortunes f_m for $m < n$. So perhaps there is (at least in every leavable Γ) a stationary family of nearly optimal strategies.

A single strategy σ is *stationary* if $\sigma[f]$, or more logically $\sigma[\cdot]$, is a stationary family. Therefore, whether a single strategy σ is stationary does not depend on σ_0.

There can be, at each f, a stationary optimal strategy and yet be no stationary family of optimal strategies.

Example 1. Let: F be the integers; γ be a diffuse gamble on the positive integers; $\Gamma(j) = \{\delta(j), \gamma, \delta(-j)\}$ for $j \geq 0$, $\Gamma(j) = \{\delta(j)\}$ for $j < 0$; $u(j) = 0$ for $j \geq 0$, $u(j) = 1 + j^{-1}$ for $j < 0$.

Four lemmas prepare for the first of several theorems on the existence of stationary families of optimal and near optimal strategies.

LEMMA 1. For any f and positive ϵ there is available in Γ at f a policy $\pi = (\sigma, t)$ such that, with positive σ-probability, f_t is ϵ-adequate and σ ϵ-conserves V at f up to t.

LEMMA 2. If F is finite, then, for some sufficiently small positive ϵ, every ϵ-adequate f is adequate. If also each $\Gamma(f)$ is finite, then ϵ can be so chosen that also, for every f, every γ in $\Gamma(f)$ that ϵ-conserves V at f conserves V at f.

LEMMA 3. Let $\bar{\sigma}$ be a stationary family of strategies. If for each positive ϵ there is a family $\bar{t}$ of stop rules for which the $\bar{\sigma}(f)$-probability that $f_{\bar{t}(f)}$ is ϵ-adequate is uniformly positive in f, then $u(\bar{\sigma}(f)) = V(\bar{\sigma}(f))$ for all f.

Proof: Let $\bar{\gamma}$ be the gamble-valued function corresponding to $\bar{\sigma}$. Apply Theorem 8.5 to the gambling problem for which $\Gamma'(f) = \{\bar{\gamma}(f)\}$ and for which u' is the indicator of the ϵ-adequate fortunes and conclude that $V' \equiv 1$ and therewith that $\bar{\sigma}(f)u' \equiv 1$. Theorem 7.2 now applies to $\bar{\sigma}(f)$ as a strategy available in Γ at f. ◆

LEMMA 4. If F is finite and at each f there is available in Γ a policy $\pi(f) = (\sigma(f), t(f))$ such that, with positive $\sigma(f)$-probability, $f_{t(f)}$ is an adequate fortune and $\sigma(f)$ conserves V at f up to $t(f)$, then there exists a stationary family of optimal strategies.

Proof: Every stop rule is bounded according to Theorem 2.9.1. If f is adequate, let $d(f)$ be 0; otherwise let $d(f)$ be the least integer such that, for some $\pi(f)$ of the postulated kind, $t(f) \leq d(f)$. From now on, let $\pi(f)$ be of the postulated kind; and let it be so chosen that $t(f) \leq d(f)$, unless f is adequate. Let $\bar{\gamma}(f)$ be the initial gamble of $\sigma(f)$; $\bar{\gamma}(f)$ plainly must conserve V at f.

The stationary family of strategies $\bar{\sigma}$ determined by $\bar{\gamma}$ is optimal. Why? If $d(f) > 0$, the $\bar{\gamma}(f)$-probability of those fortunes f' for which $d(f') < d(f)$ is positive. Since F is finite and d is bounded, there is an $\epsilon > 0$ and a family of stop rules $\bar{t}$, one stop rule for each f, such that, with $\bar{\sigma}(f)$-probability at least ϵ, $f_{\bar{t}(f)}$ is adequate. Lemma 3 now shows that each $\bar{\sigma}(f)$ equalizes u and V. Theorems 5.1 and 6.1 imply that each $\bar{\sigma}(f)$ is optimal. ◆

THEOREM 1. If F is finite and each $\Gamma(f)$ is finite, then there is a stationary family of optimal strategies.

Proof: Choose ϵ as in the second part of Lemma 2. With this ϵ, choose $\pi(f)$ as in Lemma 1. The hypotheses of Lemma 4 are then satisfied. ◆

The finiteness of both F and $\Gamma(f)$ is essential for Theorem 1, as the examples in Section 8 amply illustrate. Still more, even if F is finite, no stationary strategy available at some f may be at all good.

Example 2. Let: $F = \{0, 1\}$; $u(i) = i$ for $i = 0$ and 1; $\Gamma(0) = \{\delta(0)\}$, and $\Gamma(1)$ be the set of all γ such that, for some $\epsilon > 0$, $\gamma\{0\} = \epsilon$ and $\gamma\{1\} = 1 - \epsilon$. For any stationary σ available at 1, $u(\sigma) = 0$. But $V(1) = 1$, as is easily seen.

We believe that if Γ is leavable then there does exist a stationary family of nearly optimal strategies, but we have established this only in certain special cases.

THEOREM 2. If Γ is leavable and F is finite, then for each positive ϵ there is available in Γ a stationary family of strategies $\bar{\sigma}$ such that $u(\bar{\sigma}(f)) > U(f) - \epsilon$ for all f.

Proof: For each f, let $\pi(f)$ be available in Γ at f and such that $u(\pi(f)) \geq U(f) - \epsilon$. Since, according to Theorem 2.9.1, every stop rule is bounded, only a finite number of gambles are used by the family $\pi(f)$. Let Γ' be the smallest leavable subhouse of Γ for which $\Gamma'(f)$ includes all γ used by any $\pi(f')$ at f. Theorem 1 applied to Γ' yields a stationary family $\bar{\sigma}$ available in Γ' such that $u(\bar{\sigma}(f)) = V'(f) = U'(f) \geq U(f) - \epsilon$. Of course, $\bar{\sigma}$ is available in Γ too. ◆

Donald Ornstein has extended Theorem 2 to the situation in which F is countable and each $\gamma \in \Gamma(f)$ is countably additive but has not yet published his proof. Whether Theorem 2 can be further extended to uncountable F or to finitely additive γ is not known.

THEOREM 3. If F is finite and there is an optimal strategy at each f, then there is a stationary family of optimal strategies.

Proof: Let $\sigma(f)$ be optimal at f, and let $\epsilon > 0$ satisfy the first part of Lemma 2. Theorem 7.2 assures the existence of a stop rule $t(f)$ such that $(\sigma(f), t(f))$ terminates with positive probability at an ϵ-adequate fortune which, here, is necessarily adequate. According to Theorem 6.3, with $\sigma(f)$-probability 1, $\sigma(f)$ conserves V at f up to $t(f)$. Consequently, the hypotheses of Lemma 4 are satisfied. ♦

The finiteness of F is essential for Theorem 3. For infinite F there can be optimal strategies at every f and yet no stationary optimal strategy at any f. This can happen even if every gamble is a one-point gamble and Γ is leavable.

Example 3. Let: F be the integers; $\Gamma(j) = \{\delta(j), \delta(j+1), \delta(-j)\}$ for $j \geq 0$; $\Gamma(j) = \{\delta(j), \delta(j+1)\}$ for $j < 0$; $u(j) = 0$ for $j \geq 0$, $u(j) = 1 + j^{-1}$ for $j < 0$.

The finiteness of F in Theorem 3 cannot be relaxed even if Γ is leavable and there is an optimal policy (not merely an optimal strategy) at each f, and even if the conclusion is weakened to assert only the existence of an optimal stationary strategy at each f.

Example 4. Let F be the pairs of integers $f = (i, j)$, with $i = 0$, 1, 2, and $j \geq 1$. Let γ be a gamble that attaches probability 2^{-j} to $f = (1, j)$ and γ' be a diffuse gamble on the fortunes $(1, j)$. Let $\Gamma(0, j) = \{\delta(0, j), \delta(2, j), \gamma\}$, $\Gamma(1, j) = \{\delta(1, j), \delta(2, j), \gamma'\}$, and $\Gamma(2, j) = \{\delta(2, j)\}$. Let $u(i, j) = i - j^{-1}$. Let $\sigma_0 = \gamma$, $\sigma_1(f_1) = \gamma'$ if $i_1 = 1$ and $\delta(f_1)$ otherwise; $\sigma_2(f_1, f_2) = \delta(2, j_2)$; $\sigma_n(f_1, \cdots, f_n) = \delta(f_n)$ if $n > 2$. The strategy σ thus defined is available in Γ at $(0, j)$ and $u(\sigma) = 2 = \sup u$; so σ is optimal. In fact, optimal policies for each initial f are easily defined. But no stationary strategy in Γ at $(0, j)$ is optimal.

If there is complete absence of any stochastic element, the gambler can restrict himself to stationary families without macroscopic penalty. A house Γ is *deterministic* if each γ in every $\Gamma(f)$ is $\delta(g)$ for some g. A strategy σ available in Γ at f is *strongly optimal* at f, if $\lim u(\sigma, t)$ (not only $\lim\sup u(\sigma, t)$) is $V(f)$.

THEOREM 4. If Γ is deterministic, then the existence of a stationary family of optimal strategies is implied by each of these two conditions:

(*a*) u assumes only a finite number of values.

(*b*) There is at each f a strongly optimal strategy.

Proof: Since the two parts of this theorem have similar proofs, only that of Part (*b*) (which is, if anything, the more difficult) is given here.

Every strategy available in Γ is $\delta(h)$ for some history $h = (f_1, f_2, \cdots)$. Choose for each f_0 a strongly optimal strategy $\delta(h)$.

Suppose first that for each f' there are at most a finite number of n for which $f_n = f'$. Define a sequence $f_0^*, f_1^*, \cdots$ inductively, thus: Let $f_0^* = f_0$, and let f_{n+1}^* be f_j, where j is the smallest integer such that $f_{j+i} \neq f_n^*$ for all $i \geq 0$. As is easily verified: $f_j^* \neq f_i^*$ for $0 \leq i < j$; for each i, there is a j such that $f_i^* = f_j$ and $f_{i+1}^* = f_{j+1}$; and $\delta(h^*)$ is a stationary strategy that is strongly optimal for f_0, where $h^* = (f_1^*, f_2^*, \cdots)$.

Suppose next that there is an f' such that, for an infinite number of n, $f_n = f'$. Since $\delta(h)$ is strongly optimal, $u(f') = V(f_0)$. It requires but modest effort to define a sequence $f_0^*, f_1^*, \cdots$ with the following properties: (i) $f_0^* = f_0$; (ii) for every i there is a j such that $f_i^* = f_j$ and $f_{i+1}^* = f_{j+1}$; (iii) if $f_i^* = f_k^*$, then $f_{i+1}^* = f_{k+1}^*$; and (iv) for an infinite number of i, $f_i^* = f'$. Let $h^* = (f_1^*, f_2^*, \cdots)$. Then (i) and (ii) imply that $\delta(h^*)$ is available at f_0 since $\delta(h)$ is. According to (iii), $\delta(h^*)$ is stationary, and $\delta(h^*)$ is seen to be optimal according to (iv). Thus, there is a stationary family of optimal strategies defined at least for a nonempty subset G of F. By Zorn's lemma (Kelley 1955), there exists a maximal such family. It is easy to see that such a maximal family is necessarily defined for all of F. ◆

Since any bounded u can be approached uniformly from below by a u' with only a finite number of values, Part (*a*) of Theorem 4 has this corollary:

COROLLARY 1. If Γ is deterministic and ϵ is positive, there is available a stationary family of strategies $\bar{\sigma}$ such that $u(\bar{\sigma}(f)) \geq V(f) - \epsilon$ for all f.

Example 3 illustrates the need for some condition like (*a*) or (*b*) in Theorem 4. If Γ in Theorem 4 is also leavable, the proof is simplified, and (b) then implies the existence of a stationary family of strongly optimal strategies. But without leavability the conclusion cannot be thus strengthened, as easy examples show.

Theorem 4 can of course be interpreted as facts about directed graphs. These facts may previously have been observed in that context.

THEOREM 5. If each $\Gamma(f)$ contains at most one gamble $\alpha(f)$ in addition to $\delta(f)$, then the stationary family determined by

$$\gamma(f) = \begin{cases} \delta(f) & \text{for adequate } f \text{ if } \delta(f) \text{ is available} \\ \alpha(f) & \text{for other } f \end{cases} \tag{1}$$

is optimal.

Since this theorem is so special and will not be used in this book, we do not give a proof. There may be no optimal strategy at some f if some $\Gamma(f)$ contains as many as three elements, as is illustrated by Example 8.1.

The next theorem, though immediate from earlier results, is recorded because it is applicable to the two-armed (or n-armed) bandit problems with future incomes discounted as described in Section 12.5.

THEOREM 6. If at each f there is available a V-conserving γ, as for example when there are only a finite number of γ available at each f, and if every available strategy equalizes u and V, then there is a stationary family of optimal strategies.

in Theorem 5. If T in Theorem 5 is also [illegible], the proof simplifies [illegible]
(a) and (b) imply the existence of a stationary family [illegible] strategy
optimal strategies. But without [illegible] the conclusion [illegible]
be true [illegible] as examples show.

There [illegible] of course be interpreted as [illegible]
[illegible] These [illegible] may [illegible] have been proved [illegible]
[illegible] 4.

THEOREM 3. If each [illegible] contains at most one point [illegible] in
addition [illegible] then the stationary family determined by

(1) [illegible]

is optimal.

[illegible] is not [illegible] and [illegible] ordinary [illegible] There may be no optimal strategy at [illegible] as is illustrated by Example 8.1.

The next theorem, though [illegible] results, is needed because [illegible] problems [illegible] described in Section 12.

THEOREM 6. [illegible]

4 CASINOS WITH FIXED GOALS

1. INTRODUCTION. Almost all the specific gambling problems treated in this book involve gambling houses of a special kind, to be called casinos, and very special utility functions, called fixed goals. The problem of a casino with a fixed goal will be described and studied generally in this chapter to prepare for more specific problems in later chapters.

The gambler's fortune in a *casino* in the technical sense to be introduced here is simply the "money" available to him (in cash or credit); thus $F = [0, \infty)$. Informally, the rules that distinguish a casino from gambling houses more generally are these: The gambler cannot stake what he does not have; a rich gambler can do whatever a poor one can do, though not necessarily on a larger scale; and a poor gambler can, on a small scale, imitate a rich one.

To put these rules formally and to exploit them, it is convenient to have a compact nomenclature for the homothetic images of a gamble γ on the real numbers (not necessarily one on $[0, \infty)$). For real a and b, let $[a\gamma + b]$ be the gamble such that

$$(1) \qquad [a\gamma + b]z = \int z(af + b)\, d\gamma(f)$$

for all bounded, real-valued functions z of the real variable f; equivalently, the $[a\gamma + b]$-probability of the set $aS + b$ is the

γ-probability of S. Let $\Theta(f)$ be the set of random displacements or *lotteries* available at f. That is, $\theta \in \Theta(f)$ if and only if, for some $\gamma \in \Gamma(f)$, $\gamma = [\theta + f]$. Clearly, $\theta(-\infty, x]$ is the probability of winning at most x when using γ at the fortune f, for it equals $\gamma(-\infty, f + x]$.

The rule that the gambler cannot stake more than he possesses means that the probability of his fortune's becoming negative must be 0, and this is already implicit in the definition $F = [0, \infty)$. The rule that a rich gambler can do whatever a poor one can do means that $\Theta(tf) \subset \Theta(f)$ for $0 \leq t \leq 1$ and $0 \leq f$. The rule that the poor gambler can imitate a rich one on a small scale means that, for $0 \leq t \leq 1$ and $0 \leq f$, $[t\Theta(f)] \subset \Theta(tf)$, where $[t\Theta(f)]$ is, of course, the collection of all $[t\theta]$ for $\theta \in \Theta(f)$. For simplicity, it is also made part of the formal definition of a casino that only $\delta(0)$ is in $\Gamma(0)$.

Certain casinos as technically defined can, with a grain of salt, be taken as good models of real-world casinos. The main difference seems to be that, because of the discrete nature of real money or gambling chips, only a finite number of homothetic images of a gamble are actually available in a real-world casino.

The gambler at a casino will be said to have a *fixed goal* if his only desire is to obtain at least a certain fortune. It is often convenient to let the unit of money be this goal fortune. Henceforth, unless the contrary is explicitly stated, the gambler at a casino will be understood to have a fixed goal at 1; thus $u(f) = 0$ for $0 \leq f < 1$, and $u(f) = 1$ for $f \geq 1$.

2. THE CASINO INEQUALITY

An easily derived inequality practically characterizes the functions U for which $U(f)$ is the optimal probability of attaining the goal 1 in some casino, beginning at f. To express this inequality, and for many other purposes, it is convenient to write $\bar{x}$ for $1 - x$, where x is a number.

THEOREM 1. U defined on $[0, \infty)$ is the U of some casino (with fixed goal 1) if and only if $0 \leq U(f) \leq 1$ for $f \geq 0$, $U(0) = 0$, $U(f) = 1$ for $f \geq 1$, and

$$U(fg + \bar{f}h) \geq U(f)U(g) + \bar{U}(f)U(h) \tag{1}$$

for $0 \leq f \leq 1$ and $0 \leq h \leq g \leq 1$.

The proof is given in two parts after two easy lemmas.

LEMMA 1. In a casino, $\Theta(fg + \bar{f}h) \supset [(g - h)\Theta(f)]$ for $0 \leq f \leq 1$ and $0 \leq h \leq g \leq h + 1$.

LEMMA 2. If $0 \leq U(f) \leq 1$ for $f \geq 0$, $U(f) = 1$ for $f \geq 1$, and (1) obtains for $0 \leq f \leq 1$ and $0 \leq h \leq g \leq 1$, then U is nondecreasing in f, and (1) obtains under the more general conditions $0 \leq f$ and $0 \leq h \leq g$.

Proof of necessity: If U is the U of a casino: $0 \leq U(f) \leq 1$ because $U(f)$ is a probability; $U(0) = 0$ because only $\delta(0)$ is in $\Gamma(0)$; $U(f) = 1$ for $f \geq 1$, because if $f \geq 1$, the gambler can attain his goal by stagnating immediately.

Finally, to see the necessity of (1), argue thus: If the gambler has a fortune $fg + \bar{f}h$ between h and g with $g \geq h$, he can find a policy that will leave him at least as high as g with probability almost $U(f)$ without taking him lower than h. This is so simply because in view of Lemma 1 he can, so to speak, imitate on a small scale a policy leading from f to at least 1 with probability nearly $U(f)$. After this first phase, let the gambler then imitate a nearly optimal policy beginning at g if he finds himself as high as g, or otherwise let him imitate one beginning at h. The two phases together will leave him above 1 with at least almost the probability given by the right side of (1). The condition is therefore necessary. The argument can of course be written more formally if less clearly; see the contexts of (2.9.3) and (2.9.7).

Proof of sufficiency: The proof of sufficiency will actually show that a function which satisfies the conditions is the U of some casino such that $\Gamma(g)$ includes for every g in [0, 1] the gamble which assumes the value 1 with probability $U(g)$ and 0 with probability $\bar{U}(g)$. Call this gamble γ_g, and let $\theta_g = [\gamma_g - g]$ be the corresponding lottery. For each $f \geq 0$, let $\Theta(f)$ be the set of all lotteries $[t\theta_g]$ for all g in [0, 1] and all t such that $0 \leq tg \leq f$. Check that $\Gamma(f) = [\Theta(f) + f]$ is a casino. Now Theorem 2.12.3 applies to show that U is the U of this casino, thus: First, the value $U(f)$ can be attained by a stagnant strategy of one move; $\gamma_f u = \bar{U}(f)u(0) + U(f)u(1)$. This is $U(f)$, as it should be, since $u(1) = 1$ and $u(0) = 0$. However, for $tg \leq f$,

$$\begin{aligned}[t\gamma_g + (f - tg)]U &= U(g)U(f + t\bar{g}) + \bar{U}(g)U(f - tg) \\ &\leq U(g(f + t\bar{g}) + \bar{g}(f - tg)) = U(f).\end{aligned}$$

This inequality is given directly by the hypothesis if $f + t\bar{g} \leq 1$ and $g \leq 1$, and by Lemma 2 in all cases. ♦

The proof obviously also shows that the smallest casino such that $\theta_g \in \Theta(g)$ also has U as its utility.

In view of Theorem 1, it is easy to verify that the collection of functions $U(f)$ that are the U of some casino are a semigroup under composition. That is, if $U(f)$ and $V(f)$ are such functions, then so is $U(V(f))$.

A natural one-parameter semigroup of casino functions are the functions $\min(f^\alpha, 1)$ for $\alpha \geq 1$. The essential step to verify that these are casino functions, by means of Theorem 1, is to check the inequality

$$(2) \qquad fg + \bar{f}h \geq [f^\alpha g^\alpha + (1 - f^\alpha)h^\alpha]^{1/\alpha}$$

for f in $[0, 1]$ and $0 \leq h \leq g$. Equality obviously obtains in (2) if f is 0 or 1 or if $g = h$. If $h < g$, the right side of (2) is strictly convex. Therefore the inequality obtains, as asserted, and is strict except in the cases specified. Interesting casinos with $\min(f^\alpha, 1)$ as casino functions are mentioned at the end of Section 9.3.

3. EXPLORATION OF THE CASINO INEQUALITY

What does the casino inequality (2.1) imply about *casino functions,* that is, functions that are the U of a casino in the interesting range $0 \leq f \leq 1$?

It is convenient to explore the casino functions in terms of two inequalities simpler than (2.1) that follow from (2.1) and from the fact that 0 and 1 are fixed points for U.

$$(1) \qquad \begin{aligned} U(fg) &= U(fg + \bar{f}0) \\ &\geq U(f)U(g). \end{aligned}$$

$$(2) \qquad \begin{aligned} \bar{U}(\overline{fg}) &= 1 - U(1 - fg) \\ &= 1 - U(\bar{f}1 + f\bar{g}) \\ &\leq 1 - U(\bar{f}) - \bar{U}(\bar{f})U(\bar{g}) \\ &= \bar{U}(\bar{f})\bar{U}(\bar{g}). \end{aligned}$$

The final form of (2) looks especially pretty with (1), but the following form is also useful.

$$U(f + g - fg) \geq U(f) + U(g) - U(f)U(g). \tag{3}$$

Though (1) and (2) do not imply the casino inequality, as will be shown in Section 8, they lead to several interesting properties of casinos. Anything true of all bounded, nonvanishing solutions of (1) and (2) is *a fortiori* true of the U of any casino.

Inequalities (1) and (2) point to the answer to an incidental question: "For what casinos does equality hold in every instance of the casino inequality (2.1)?" For the U of such a casino, equality would hold throughout in (1) and (2). It is not difficult to verify that the only bounded solutions in the interval [0, 1] of the identity corresponding to (1) are the constants 0 and 1, the indicator functions of (0, 1] and of $\{1\}$, and functions of the form f^α, $\alpha > 0$. The same reasoning applied to (3) requires $\bar{U}$ to be one of the same four indicator functions or $\bar{f}^\beta$, $\beta > 0$. The functions compatible with both constraints are only the four indicator functions and f. Of these, all but the constants 0 and 1 are casino functions. The function 0 pertains to no casino since it is not 1 at 1; the function 1 pertains to no casino since it is not 0 at 0.

Inequalities (1) and (2) separately imply, for every positive integer n,

$$U(f^n) \geq U^n(f), \tag{4}$$

$$\bar{U}(\overline{f^n}) \leq (\bar{U}(\bar{f}))^n. \tag{5}$$

The case $n = 2$ shows that $U(f) \geq 0$ for all f, that is, for all f in the range of interest [0, 1]. Also $U^2(0) \leq U(0)$, $\bar{U}(0) \leq \bar{U}^2(0)$; so $U(0) = 0$ or 1, and similarly $U(1) = 0$ or 1.

If $U(1) = 0$, then, according to (3), $0 = U(1) = U(f + 1 - f \cdot 1) \geq U(f) \geq 0$, so that U is identically 0. The function 0 is of course a solution of (1) and (2) (and even of (2.1)), though it pertains to no casino.

If $U(0) = 1$, then, again according to (3), $U(f) = U(f + 0 - f \cdot 0) \geq U(f) + U(0) - U(f)U(0) = 1$; so $U(f) \geq 1$. One such solution is the function 1. Obviously none could pertain to any casino. All solutions of (1) and (2) that exceed 1 somewhere are necessarily unbounded near $f = 0$, as (4) shows. There are unbounded solutions of (1) and (2). For example, let $U(0) = 0$ and $U(f) = f^{-1}$ for $f > 0$; then $U(fg) = U(f)U(g)$ for $f, g \geq 0$, and $\bar{U}(f) \leq 0$ for $1 \geq f > 0$, which makes (1) and (2) evident. This function also satisfies (2.1).

To sum up, the bounded solutions U of (1) and (2) are confined between 0 and 1. Apart from the two exceptional ones $U \equiv 0$ and $U \equiv 1$, they all satisfy $U(0) = 0$ and $U(1) = 1$. From now until further notice, assume that U is a bounded solution of (1) and (2), with $U(0) = 0$ and $U(1) = 1$.

Such a U is nondecreasing in f. In fact, if $0 < g \leq f + g \leq 1$, then $g/\bar{f} \leq 1$, and, according to (3),

$$(6) \qquad \begin{aligned} U(f + g) &= U\left(f + \frac{g}{\bar{f}} - f\frac{g}{\bar{f}}\right) \\ &\geq U(f) + U\left(\frac{g}{\bar{f}}\right) - U(f)U\left(\frac{g}{\bar{f}}\right) \\ &= U(f) + \bar{U}(f)U\left(\frac{g}{\bar{f}}\right). \end{aligned}$$

This not only proves the required monotony but shows that it is strict at f from the right unless $U(f) = 1$ or $U(h) = 0$ for some positive h. These exceptions can occur, but only in very special ways. First, if $U(f) = 1$ for some f with $0 < f < 1$, then (4) shows immediately that $U(f) = 1$ for some arbitrarily small f, and therefore $U(f) = 1$ for all positive f in view of the demonstrated monotony. If $U(h) = 0$ for some positive h, then (4) applied to $h^{1/n}$ shows that $U(f) = 0$ for all $f < 1$. Thus U is either 1 in (0, 1) or 0 there (which are real possibilities), or else it is strictly increasing throughout [0, 1]. (Incidentally, (1) cannot be used instead of (3) to prove monotony here.) Assume from now until further notice that U is strictly increasing.

Under the assumptions now in force, U is demonstrably continuous. First, U is continuous at 1 because, for any f in (0, 1), (5) is impossible unless $\bar{U}(\overline{f^n})$ approaches 0 as required. Substituting $f^{1/n}$ for f in (5) shows that U is also continuous at 0. If $f > g > 0$, then

$$(7) \quad 0 \leq U(f) - U(g) = U(f) - U\left(f\frac{g}{f}\right) \leq U(f)\left[1 - U\left(\frac{g}{f}\right)\right],$$

which proves continuity in view of the established continuity at 0 and 1.

As will be shown next, $U(f) > f$ is impossible for a strictly increasing solution of (1) and (2). Let g maximize $U(f) - f$, and suppose $U(g) = \alpha g > g$. Then $U(g^n) \geq \alpha^n g^n$, and

$$U(g + \bar{g}g^n) - (g + \bar{g}g^n) \geq [U(g) - g] + \bar{U}(g)U(g^n) - \bar{g}g^n$$
$$\geq [U(g) - g] + [\bar{U}(g)\alpha^n - \bar{g}]g^n.$$

Since $\bar{U}(g) > 0$ and $g > 0$, the left member of the sequence of inequalities exceeds $U(g) - g$ for large n, which is a contradiction.

The possibility $U(g) = g$ for some g, $0 < g < 1$, is not excluded, but if it occurs once, it occurs everywhere, as will be shown. This possibility does pertain to a casino, according to Theorem 2.1; a specific example is the casino in which all fair, and none but fair, gambles are permitted, as is not hard to see. If $U(g) = g$, then $U(g^n) \geq g^n$; but inequality here has been shown to be impossible, so $U(g^n) = g^n$. Dually, $\bar{U}(\overline{(\bar{g}^n)}) \leq (\bar{U}(g))^n = \bar{g}^n$; so $U(1 - \bar{g}^n) = 1 - \bar{g}^n$. Now, $U(f) = f$ for every f of the form $1 - (1 - g^m)^n$. These numbers are everywhere dense in $[0, 1]$, and since U is continuous, the assertion is proved.

Once attention is focused on strictly increasing functions, which has been justified by study of (1) and (2) without the stronger inequality (2.1), the casino inequality (2.1) and its consequences (1) and (2) can be couched in a particularly graphic and suggestive form.

$$\frac{U(fg + \bar{f}h) - U(h)}{U(g) - U(h)} \geq U(f) \tag{8}$$

for $0 \leq f \leq 1$ and $0 \leq h < g \leq 1$.

$$\frac{U(fg)}{U(g)} \geq U(f) \tag{9}$$

and

$$\frac{U(f + \bar{f}g) - U(g)}{1 - U(g)} \geq U(f) \tag{10}$$

for $0 \leq f \leq 1$, $0 < g < 1$. Inequality (8) says that, if any segment of the graph of U is linearly transformed horizontally and vertically so as to be 0 at $f = 0$ and 1 at $f = 1$, then the transformed segment never descends below the graph of U; (9) and (10) say the same, but only for segments beginning at 0 and for segments ending at 1, respectively.

With this graphical interpretation, it is particularly easy to see that the inequalities put certain constraints on the derivatives of U. To express these, let U^* and U_*, ${}^*U$ and ${}_*U$ denote the upper and lower right, and upper and lower left, derivatives of U. Then (8) implies two pairs of inequalities:

$$(11) \qquad \left.\begin{matrix} {}^*U(g) \\ {}_*U(g) \end{matrix}\right\} \leq \frac{U(g) - U(h)}{g - h} \times \left\{\begin{matrix} {}^*U(1) \\ {}_*U(1) \end{matrix}\right.$$

and

$$(12) \qquad \left.\begin{matrix} U^*(h) \\ U_*(h) \end{matrix}\right\} \geq \frac{U(g) - U(h)}{g - h} \times \left\{\begin{matrix} U^*(0) \\ U_*(0) \end{matrix}\right.$$

for $0 \leq h < g \leq 1$.

Setting $h = 0$ in (11) and $g = 1$ in (12) gives two pairs of inequalities that are implied separately by (9) and (10) and, therefore, by (1) and by (2). Letting h approach g in the upper part of (11) and g approach h in the lower part of (12) leads to the inequalities ${}_*U(g)/{}^*U(g) \geq 1/{}^*U(1)$ and $U_*(g)/U^*(g) \geq U_*(0)$. (Letting h approach g in the lower part of (11) and g approach h in the upper part of (12) leads only to the conclusion that $U^*(0) \leq 1 \leq {}_*U(1)$, which has already been proved in the stronger assertion that $U(f) \leq f$ for all f.) Possibly more than (11) and (12) can be inferred about the regularity of U, but examples in later chapters will show that U can, for instance, be a singular monotonic function, and so it is dangerous to expect too much regularity. On the other hand, it is not unusual for U to be analytic, as the functions f^α for $\alpha \geq 1$, and other examples in later sections, illustrate.

An illustrative application of (12) was brought to our attention by Lawrence Jackson and Roger Purves. For what values of α greater than 1 and of λ strictly between 0 and 1 is the convex combination $\lambda f + \bar{\lambda} f^\alpha$ of the two casino functions f and f^α itself a casino function? Part of the answer follows on applying (12) with $g = 1$ to conclude that

$$(13) \qquad \lambda + \bar{\lambda}\alpha h^{\alpha-1} \geq \frac{1 - (\lambda h + \bar{\lambda} h^\alpha)}{1 - h}\,\lambda$$

if $\lambda f + \bar{\lambda} f^\alpha$ is a casino function; for (13) is false for small h if α exceeds 2.

Jackson has shown that $\lambda f + \bar{\lambda} f^\alpha$ is a casino function for $1 \leq \alpha \leq 2$. His result is proved as Theorem 8.5.6.

The first special casino inequality, $U(fg) \geq U(f)U(g)$, can be viewed as a multiplicative expression of subadditivity. This leads to the information that U approximates a monomial near the origin. Let $V(z) = -\log U(e^{-z})$ for nonnegative z. Then, V is continuous; $z \leq V(z)$ and $V(z + w) \leq V(z) + V(w)$ for all nonnegative z and w. According to a theorem on subadditive functions (see, for example, (Hille and Phillips 1947, Theorem 7.6.1, p. 244)), $V(z)/z$ approaches α as z approaches ∞, where $1 \leq \alpha = \inf (V(z)/z)$. This means that

$$f^{\alpha+\epsilon} < U(f) \leq f^\alpha, \tag{14}$$

where the first inequality holds for all positive f less than some positive f_0 which depends on the positive number ϵ, and the second inequality holds for all f in [0, 1]. Of course,

$$1 \leq \alpha = \inf \frac{\log U(f)}{\log f} \qquad \text{for } f \text{ in } (0, 1). \tag{15}$$

From a somewhat similar use of (2), or from a duality principle spelled out in Section 6, the dual conclusions follow. That is, for some unique β in (0, 1]

$$\bar{f}^\beta \leq \bar{U}(f) < \bar{f}^{\beta-\epsilon}, \tag{16}$$

where the first inequality holds for all f in [0, 1] and the second holds for all f in $(f_1, 1)$, where f_1 depends on the positive number ϵ. For later reference,

$$1 \geq \beta = \sup \frac{\log \bar{U}(f)}{\log \bar{f}} \qquad \text{for } f \text{ in } (0, 1). \tag{17}$$

Whether the full casino inequality implies much more about the behavior of U near 0 and 1 we do not know. But it seems doubtful, for example, that the strict inequality in (14) can be strengthened to $kf^{\alpha} < U(f)$ for some positive k.

4. THE THREE SPECIAL TYPES OF CASINOS

Some of the conclusions of this chapter thus far can be put as follows. There are three special U's that pertain to casinos: U and its casino are *trivial*, *fair*, or *superfair* according as $U(f) = 0, f$, or 1 for f in (0, 1). All other U's that pertain to some casino rise strictly, monotonically, and continuously, from the point (0, 0) to (1, 1) with $U(f) < f$ for f in (0, 1), and subject to the inequalities (3.8) through (3.12), and are the U's of the *subfair* casinos. Thus casinos fall into three special types and a general type.

What are the conditions on the lotteries available in a casino that determine to which of the types the casino belongs? Much of the answer can be presented immediately. However, the full distinction between fair and subfair casinos remains for Chapter 11.

Trivial casinos. A casino is trivial if and only if all probability of macroscopic gain is excluded, that is, $\theta(\epsilon, \infty) = 0$ for all positive ϵ and all θ in $\Theta(1)$ (equivalently, for all θ in some $\Theta(f)$ for $f > 0$).

Proof: The condition is obviously necessary; it is sufficient in view of the basic Theorem 2.12.1. ♦

Superfair casinos. A casino is superfair if and only if $\Theta(1)$ (or, equivalently, $\Theta(f)$ for all $f > 0$) contains a *superfair lottery* θ, that is, one for which $\int \min(x, k)\, d\theta(x) > 0$ for some positive real number k.

Proof: The condition is necessary. For without it, the function $Q(f) = \min(f, 1)$ majorizes U according to the basic Theorem 2.12.1.

Suppose now that $\Theta(1)$ (and therefore every $\Theta(f)$ for $f > 0$) contains superfair lotteries, but that Γ is not superfair. A contradiction arises thus. Since Γ is not trivial, U is continuous and strictly increasing, and $U(f) \leq f$. Consequently, U restricted to [0, 1] is supported from below by some line of slope 1; thus there is a smallest nonnegative number α such that $U(f) \geq f - \alpha$ for all f in [0, 1], and a g in (0, 1) such that $U(g) = g - \alpha$. If θ is a superfair lottery in $\Theta(1)$, $[t\theta]$ is in $\Theta(g)$ for $0 < t < g$, and, for sufficiently small t, U fails to be excessive for $[g + t\theta]$ according to the following calculation.

$$(1)\qquad \begin{aligned}[g + t\theta]U &= \int U(g + z)\, d[t\theta](z) \\ &\geq \int \min\,(g + z - \alpha, 1)\, d[t\theta](z) \\ &= g - \alpha + \int \min\,(z, 1 - g + \alpha)\, d[t\theta](z) \\ &= U(g) + \int \min\,(z, \bar{U}(g))\, d[t\theta](z) \\ &= U(g) + t \int \min\left(z, \frac{\bar{U}(g)}{t}\right) d\theta(z) \\ &> U(g)\end{aligned}$$

for small positive t. This contradicts Theorem 2.14.1. ◆

Fair casinos. Of course, for Γ to be fair it is necessary that $\Theta(1)$ contain no superfair lotteries. In the presence of this necessary condition, it is natural to expect that Γ is fair if $\Theta(1)$ contains a *fair lottery*, that is, one for which

$$(2)\qquad \lim_{k \to \infty} \int \min\,(x, k)\, d\theta(x) = 0.$$

This condition is not necessary; for clearly Γ can be fair if $\Theta(1)$ contains no fair gambles but does contain arbitrarily good facsimiles of them. Of greater interest, $U(f)$ can be strictly less than f in $(0, 1)$ in the presence of a fair, *nontrivial* lottery, that is, one for which $\theta(-\epsilon, \epsilon) < 1$ for some $\epsilon > 0$ (shown in Section 11.2). As is to be expected, if a fair, nontrivial θ has bounded support so that $\theta(k, \infty) = 0$ for some $k > 0$, every casino containing θ is at least fair. A technique showing that some fair θ with unbounded support also generate fair casinos will be illustrated here, though this technique will be largely superseded in Section 11.3.

It will be shown that if (2) is replaced by the stronger condition

$$(3)\qquad \lim_{k \to \infty} \frac{\int \min\,(f, k)\, d\theta(f)}{k\theta(k, \infty)} = 0,$$

then the casino is fair. Since (3) is equivalent to

$$\lim_{k \to \infty} \frac{-1}{k\theta(k, \infty)} \int_{f \le k} f \, d\theta(f) = 1, \tag{4}$$

(3) says that the conditional expectation of the gambler's gain, given that it exceeds k, is $k + o(k)$. If, for example, $\theta(k, \infty) = e^{-k-1}$ for k in $(-1, \infty)$, then the fraction in (3) is k^{-1}.

Proof: Suppose that $U(f) < f$ for f in (0, 1) for a casino that includes a θ satisfying (3). Let α and g be as in the discussion of superfair casinos, and calculate much as in (1), but taking advantage of (4) and of the fact that $\bar{U}(g) > \bar{g}$.

$$\begin{aligned}
(5) \quad [g + t\theta]U &= \int U(g + z) \, d[t\theta](z) \\
&= \int_{z \le \bar{g}} U(g + z) \, d[t\theta](z) + [t\theta](\bar{g}, \infty) \\
&\ge \int_{z \le \bar{g}} (U(g) + z) \, d[t\theta](z) + [t\theta](\bar{g}, \infty) \\
&= U(g) + \int_{z \le \bar{g}} z \, d[t\theta](z) + \bar{U}(g)[t\theta](\bar{g}, \infty) \\
&= U(g) + t \int_{z \le \bar{g}/t} z \, d\theta(z) + \bar{U}(g)\theta\left(\frac{\bar{g}}{t}, \infty\right) \\
&= U(g) + t \left\{ \int_{z \le \bar{g}/t} z \, d\theta(z) + \frac{\bar{U}(g)}{t} \theta\left(\frac{\bar{g}}{t}, \infty\right) \right\} \\
&= U(g) + t \left\{ \int_{z \le \bar{g}/t} z \, d\theta(z) + \frac{\bar{U}(g)}{\bar{g}} \left(\frac{\bar{g}}{t}\right) \theta\left(\frac{\bar{g}}{t}, \infty\right) \right\} \\
&> U(g)
\end{aligned}$$

for small positive t. But this contradicts Theorem 2.14.1, as before. ◆

The U of a casino Γ is not necessarily determined by $\Theta(1)$ (or any other $\Theta(f)$) alone. But whether Γ is of one of the three special types—and if so, of which type it is—is determined by $\Theta(f)$ for any positive f. So far as trivial and superfair types are concerned, this fact is now evident. For fair casinos it is less evident and will be shown in Section 11.3.

5. HOW NICE CASINOS ARE

We have been working throughout at such a level of generality that the gambles γ may be only finitely additive. On this account it must be asked whether bizarre things do not happen. Are bounded stop rules nearly adequate, or can U_ω really be less than U? To what extent do the probabilities that the gambles γ assign to the non-Borel sets affect the value of U? Theorem 1 and Corollary 1 show that with respect to these questions casinos behave quite nicely.

LEMMA 1. The casino inequality holds for U_ω.

Proof: Same as for U. ◆

An *interval* is any connected set of real numbers.

LEMMA 2. If ϕ is a bounded monotone function and γ is any (finitely additive) probability measure defined on subsets of the real line, then $\gamma\phi$ is determined by the values that γ assumes on the intervals.

LEMMA 3. Each U_m of any casino is a nondecreasing function of f.

Proof: Use the hypothesis that a rich man can do whatever a poor one can, and the fact that $u = U_0$ is monotone nondecreasing. Proceed by induction, using Theorem 2.15.2. ◆

THEOREM 1. In every casino, $U_\omega = U$.

Proof: Lemma 1 implies that U_ω is continuous or is one of the two discontinuous casino functions of Section 4. If U_ω is continuous, use Lemma 3 and apply Theorem 2.15.5, part (*e*). If $U_\omega(f) = 0$ for $0 \leq f < 1$, then all probability of a macroscopic gain is excluded. Use Theorem 2.12.1 to conclude that $U(f) = 0$ for $0 \leq f < 1$. Lastly, suppose that $U_\omega(f) = 1$ for $0 < f \leq 1$. Since in all houses $U_\omega \leq U$, and no U exceeds this U_ω, $U_\omega = U$. ◆

Theorem 1 implies that a gambler with a fixed goal in a casino can restrict himself to bounded stop rules without suffering appreciable loss.

Two gambles γ and γ' are *companions* if $\gamma\phi = \gamma'\phi$ for all bounded continuous ϕ; two houses Γ and Γ' are *companions* if, for all f and every γ in $\Gamma(f)$, there is a companion of γ in $\Gamma'(f)$, and vice versa. To say that U is determined by the values that the gambles in Γ assign to continuous functions is to say that $U = U'$ for every companion Γ' of Γ.

THEOREM 2. In a casino, U is determined by the values that the gambles assign to continuous functions on $[0, \infty)$.

Proof: If U and the U' of a companion Γ' of Γ are continuous, then, for $\gamma \in \Gamma(f)$, $\gamma U' = \gamma' U' \leq U'(f)$ for some γ' in $\Gamma'(f)$, according to Theorem 2.14.1. Thus U' is excessive for Γ so that, according to the basic Theorem 2.12.1, $U' \geq U$. Similarly, $U' \leq U$; so $U' = U$. Suppose next that U is one of the two discontinuous casino functions. If U is superfair, there is, according to Section 4, a θ in $\Theta(1)$ and a positive k for which $\int \min(f, k)\, d\theta(f) > 0$. Since $\min(f, k)$ is continuous in f, there is a companion of θ, say θ', in $\Theta'(1)$ such that $\int \min(f, k)\, d\theta'(f) > 0$, which, again according to Section 4, implies that $U'(f) = 1$ for $f > 0$. A similar argument shows that, if U is trivial, so is U'. ◆

COROLLARY 1. The U of a casino is determined by the probabilities that the gambles assign to the intervals.

It is almost routine to introduce notions of convergence for gambling houses, and it is natural to inquire to what extent U varies continuously with Γ and u. Even for casinos with the usual u there is not full continuity; for a sequence of superfair casinos can converge to a fair one. This may, however, be the only source of discontinuity for casinos with the usual u.

6. THE CASINO-FUNCTION SEMIGROUP

As was mentioned after the proof of Theorem 2.1, casino functions form a semigroup with respect to substitution. To study this semigroup, let UV, in this section, denote the function for which $(UV)(f) = U(V(f))$. The two casino functions that are not strictly increasing are of little interest for the rest of this section; so usually let "casino function" mean "strictly increasing casino function", which does not interfere with the semigroup property.

If the graph of a U in the casino semigroup is regarded also as the graph of a function of the variable $1 - U$ into $1 - f$ (by looking at it sideways from the rear, so to speak), the new function is also a casino function, as the discussion of (3.8) brings out. Analytically, if U is in the semigroup, so is U^*, where

$$U^*(f) = 1 - U^{-1}(1 - f). \tag{1}$$

Since

$$U^*(1 - U(1 - f)) = 1 - U^{-1}(U(1 - f)) = f, \tag{2}$$

$$(U^*)^{-1}(f) = 1 - U(1 - f) \tag{3}$$

or $(U^*)^* = U$. Accordingly, U^* is called the *dual* of U.

What is $(UV)^*$?

$$\begin{aligned} (UV)^*(f) &= 1 - (UV)^{-1}(1 - f) \\ &= 1 - V^{-1}(U^{-1}(1 - f)) \\ &= 1 - V^{-1}[1 - (1 - U^{-1}(1 - f))] \\ &= V^*(U^*(f)); \end{aligned} \tag{4}$$

so $(UV)^* = V^*U^*$, or the duality operation is an anti-automorphism. In particular, UU^* and U^*U are self-dual. Similarly, the dual of any strictly increasing solution of either of the special inequalities (3.1) or (3.2) solves the other. Thus the intersection of the two semigroups of strictly increasing solutions of these inequalities is a semigroup that is closed under the duality operation *.

Composition and the duality operation are not the only ways to construct new casino functions out of old ones, as the next theorem implies.

THEOREM 1. The infimum of casino functions is a casino function.

Proof: Suppose that $U = \inf U_\alpha$ and that

$$U_\alpha(fg + \bar{f}h) \geq U_\alpha(f)U_\alpha(g) + \bar{U}_\alpha(f)U_\alpha(h).$$

Then $U(fg + \bar{f}h) = \inf U_\alpha(fg + \bar{f}h) \geq \inf [U_\alpha(f)U_\alpha(g) + \bar{U}_\alpha(f)U_\alpha(h)]$ $\geq \inf [U_\alpha(f)U(g) + \bar{U}_\alpha(f)U(h)] \geq U(f)U(g) + \bar{U}(f)U(h)$. ♦

COROLLARY 1. If a collection of casino functions is bounded away from 0 at a single point of the open interval (0, 1), then it is bounded away from 0 at every point in (0, 1).

(Corollary 1 is sharpened by Theorem 6.7.1.)

Since the minimum of two casino functions is a casino function, it may be interesting to observe that the operation of forming the minimum commutes with the operations of right and left composition: $U \min (V, W) = \min (UV, UW)$; $\min (V, W)U = \min (VU, WU)$. In the typical case when the functions are strictly increasing, so that the duality operation is meaningful, it, too, commutes with the operation of forming the minimum; $\min (U, V)^* = \min (U^*, V^*)$. If U^* is chosen for V, it is clear that $\min (U, U^*)$ is self-dual. Thus each casino function U immediately determines three (typically distinct) self-dual casino functions UU^*, U^*U, and $\min (U, U^*)$.

Many of the remarks above are very general and apply if U and V are arbitrary order-preserving homeomorphisms of the unit interval onto itself.

Starting with the monomials $U_\alpha(f) = f^\alpha$, which are casino functions for $\alpha \geq 1$, interesting examples of other casino functions and of relations among them are easily derived: $U_\alpha U_\beta = U_{\alpha\beta}$; $U_\alpha^*(f) = 1 - (1 - f)^{1/\alpha}$; $U_\alpha^* U_\beta^* = U_{\alpha\beta}^*$; $(U_\alpha^* U_\beta)(f) = 1 - (1 - f^\beta)^{1/\alpha}$; $U_\beta U_\alpha^* = [1 - (1 - f)^{1/\alpha}]^\beta$.

In particular, the self-dual functions $U_\alpha^* U_\alpha$ and $U_\alpha U_\alpha^*$ are different from each other if $\alpha > 1$. The point (f, W) is on the graph of the former if and only if

$$f^\alpha + (1 - W)^\alpha = 1 \tag{5}$$

and on the graph of the latter if and only if

$$(1 - f)^{1/\alpha} + W^{1/\alpha} = 1. \tag{6}$$

For all $\alpha > 1$, the first exceeds the second, at least for small f. For $\alpha = 2$, the first graph is a circular arc, and the second is a parabolic one. The third natural self-dual casino function associated with U_α, namely, $\min(U_\alpha, U_\alpha^*)$, is different from the other two since it is not everywhere analytic.

In Sections 9.2 and 9.4, a neat, self-dual, one-parameter semigroup arises naturally. It consists of the functions V_α, $\alpha \geq 1$, where

$$V_\alpha(f) = \frac{f}{\alpha - (\alpha - 1)f}. \tag{7}$$

It is easy to verify directly that $V_\alpha = V_\alpha^*$ and $V_{\alpha\beta} = V_\alpha V_\beta$. It is not difficult to give a direct proof that the V_α are casino functions, and indirect ones are in Sections 9.2 and 9.4.

It sometimes happens, as later chapters illustrate, that, when an interesting casino has U as its casino function, some other casino naturally suggested by the first one has the dual U^* as its casino function. It would therefore be interesting to formulate a concept of the dual of a casino, not merely the dual of a casino function, and even a concept of the product and infimum of two casinos corresponding to the composition and infimum of their casino functions. Our efforts on these three problems have not met with success.

If Γ and Γ^* are casinos with dual utilities U and U^* different from each other, then each is genuinely preferable to the other for some initial fortunes; that is, there are f and g such that $U(f) > U^*(f)$ and $U^*(g) > U(g)$; for the areas under U and U^* are the same.

There may be some interest in examples of functions that satisfy the casino inequality (2.1) but do not coincide in [0, 1] with the U of any casino: constant functions, linear functions $U(f) = \alpha + \beta f$ with $-\alpha\beta$ and $(1 - \alpha - \beta)\beta$ both nonnegative (except that $\alpha = 0$ and $\beta = 1$ yields a casino function), any strictly decreasing function that is never exceeded by 1, and $U(f) = f^{-1}$ for $0 < f < 1$ and $U(0) = 0$.

7. RICH MAN'S, POOR MAN'S, AND INCLUSIVE CASINOS

In any casino, a gambler can imitate a richer one on a small scale; casinos in which he can also imitate a poorer one on a large scale, so that the amounts he is allowed to stake are proportional to his wealth, are of special interest. Formally, a *rich man's* casino is one in which if the lottery θ is available at f then $t\theta$ is available at tf for all $t > 0$. In a rich man's casino, $U(f)$ is the supremum of the probability not only of attaining the goal 1 from the initial fortune f but also of attaining the goal f^{-1} times any initial fortune.

THEOREM 1. These conditions on a casino Γ are equivalent:

(*a*) Γ is a rich man's casino.

(*b*) $\Theta(f) = [f\Theta(1)]$.

(*c*) $\Theta(tf) = [t\Theta(f)]$ for $t \geq 0$.

(*d*) $[t\Theta(f)] \subset \Theta(tf)$ for $t \geq 0$.

(*e*) $[t\Theta(f)] \subset \Theta(tf)$ for $t > 1$.

Real-world casinos do not seem to be rich man's casinos. On the other hand, a real-world casino, by and large, offers to a poor man any bet offered to a rich man provided that the poor one can afford it. In a *poor man's* casino, if $0 < f < f'$, $\theta \in \Theta(f')$, and $\theta(-\infty, -f) = 0$, then $\theta \in \Theta(f)$. Trivially, for every $f > 0$ and $x > 0$, every strategy σ available at $f + x$, when translated by $-x$, is available at f, provided that σ always uses gambles attaching probability 1 to (x, ∞). Unless $\Theta(1)$ contains a θ other than $\delta(0)$ for which $\theta[0, \infty) = 1$, it is enough of a proviso that σ use only gambles attaching probability 1 to the interval $[x, \infty)$.

Because the poor man in a poor man's casino can translate the modest strategies of a richer man as described, the probability with which a gambler can increase his fortune from $f + x$ to $g + x$ without ever falling below $h + x$ is independent of x for $0 \leq h < h + x \leq f + x \leq g + x$. If $\theta[0, \infty) = 1$ only when θ is $\delta(0)$, the same argument applies even if $x = 0$. (As a matter of fact, the argument can be modified to cover $x = 0$ for every poor man's casino.)

Casinos in which a gambler may stake any amount that he possesses on any game played at the casino are of particular mathematical simplicity. An *inclusive* casino is a casino such that, for all positive t, f, and g, if $\theta \in \Theta(f)$ and $[t\theta](-\infty, -g) = 0$, then $[t\theta] \in \Theta(g)$. A necessary and sufficient condition for a casino to be inclusive is that it be both a rich man's and poor man's casino, as is easily verified. The proof of Theorem 2.1 shows that every casino function is the U of an inclusive casino. Associated with every casino Γ there are the smallest rich man's, poor man's, and inclusive casinos that include Γ.

Casinos in which all the lotteries in $\Theta(1)$ are generated from a single element θ by dilatation so that every element of $\Theta(1)$ is of the form $[t\theta]$ for some nonnegative t deserve special mention. It is easy to construct such *one-game casinos* that are neither rich man's nor poor man's casinos.

We do not know whether every casino function is the casino function of a one-game casino, though that seems unlikely to us; the minimum of two one-game casino functions is a candidate for a counterexample. If counterexamples exist, what is the smaller class?

The next two chapters are about certain particularly interesting one-game casinos.

8. THE SPECIAL CASINO INEQUALITIES DO NOT IMPLY THE GENERAL ONE.

A continuous function rising from 0 to 1 in [0, 1] can satisfy the two special casino inequalities (3.1) and (3.2) and yet fail to be a casino function. This fact will not be relied on in future arguments, so this section can be skipped or deferred.

Let $U(f) = \min (V(f), W(f))$, where

$$V(f) = 1 - (1 - f^2)^{1/2}, \tag{1}$$

$$W(f) = \bar{g} + \max (\tfrac{1}{3}(f - g), 3(f - g)). \tag{2}$$

Note that V and W, and therefore U, are self-dual in the sense that, if (x, y) is on the graph of one of them, so is $(\bar{y}, \bar{x})$.

As will be shown, if g is greater than but sufficiently close to $2^{-1/2}$, then U is not a casino function but does satisfy (3.1) and (3.2).

Assume once and for all that $2^{-1/2} < g < 3/4\ (= 9/16)^{1/2})$. It follows that $W(0)$ (which is $1 - (4g/3))$ is positive and that $U(f) = W(f)$ in and only in an interval $[a, b]$ with $a < g < b$. Consequently, $U(a) = V(a) = W(a) = \bar{b}$, and $U(b) = V(b) = W(b) = \bar{a}$. The graph of U between a and b coincides with that of W. Since $W(b) - W(a) = \bar{a} - \bar{b} = b - a$, this piece of graph, when stretched horizontally and vertically so as to connect (0, 0) with (1, 1), consists of two linear segments of slope 1/3 and 3 and therefore lies uniformly below W. Thus U fails the geometric test for a casino function.

Since U is self-dual, it remains only to check that $U(xy) \geq$

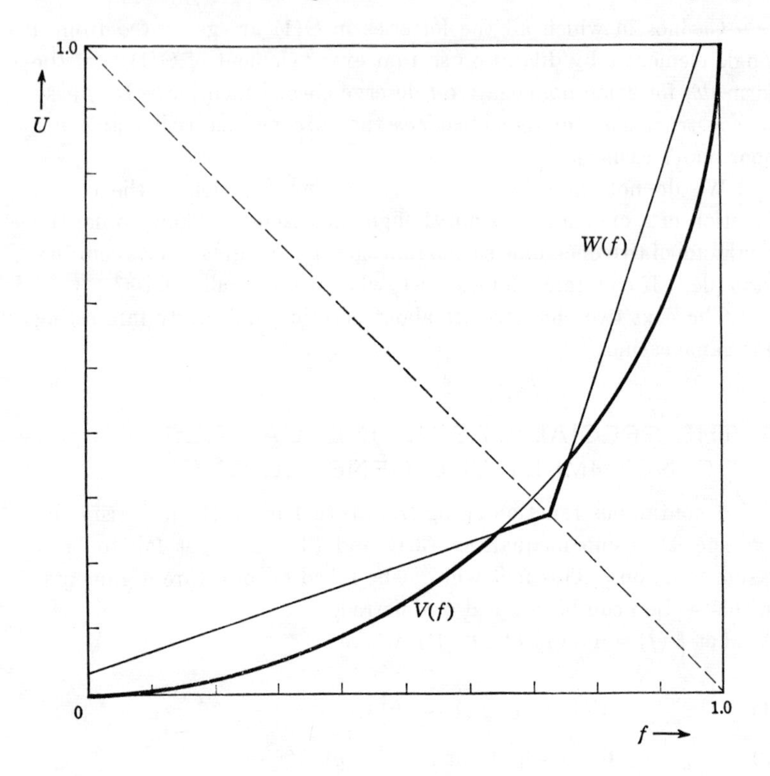

Figure 1 $U(f) = \min (V(f), W(f)); g = .73.$

$U(x)U(y)$ for all x and y in $[0, 1]$. If xy is not in $[a, b]$, then $U(xy) = V(xy) \geq V(x)V(y) \geq U(x)U(y)$; for V is a casino function and never less than U.

There is now no loss in confining the investigation to values of x and y with $x \leq y$, x in some interval $[x_1, x_2]$ with x_1 and x_2 positive. Suppose first that $y = 1 - \epsilon$ is very close to 1. Then $U(xy) = U(x(1 - \epsilon)) = U(x)(1 + O(\epsilon))$ uniformly in $[x_1, x_2]$ and in the value of g yet to be chosen. But $U(y) = U(1 - \epsilon) = 1 - [1 - (1 - \epsilon)^2]^{1/2} = 1 - (2\epsilon)^{1/2} + O(\epsilon^{3/2})$. Therefore, $U(xy) \geq U(x)U(y)$ for y in $(1 - \epsilon, 1)$.

Finally, suppose x in $[x_1, x_2]$ and y in $[x_1, 1 - \epsilon]$. On this compact set, $V(xy) - V(x)V(y) \geq \delta$ for some positive δ. If now g is so chosen that $V - U \leq \delta$, $U(xy) \geq V(xy) - \delta \geq V(x)V(y) \geq U(x)U(y)$, and the check is complete.

5
RED-AND-BLACK

1. INTRODUCTION. This chapter is mainly about a very special kind of casino, where the only game offered is that of *red-and-black.* The gambler can stake any amount s in his possession. He wins back his stake and as much more again with probability w, and he loses his stake with probability $\bar{w}$, $0 < w < 1$. Formally, if ρ is the measure that attaches probability w to 1 and $\bar{w}$ to -1, so that $[s\rho + f]$ attaches probability w to $f + s$ and $\bar{w}$ to $f - s$, then $\Gamma(f) = \{[s\rho + f]: 0 \leq s \leq f\}$; equivalently, $\Theta(f) = \{[s\rho]: 0 \leq s \leq f\}$.

The possibilities $w = 0$ and $w = 1$ could be included, but they are trivial and uninteresting. In the terminology of Section 4.4, the casino is easily seen to be superfair, fair, or subfair (but not trivial) according as w exceeds, equals, or is exceeded by 1/2; correspondingly, in (0, 1), $U(f)$ is 1, f, or positive but less than f. Attention will be focused on the subfair possibility $w < 1/2$.

It is natural to generalize ρ to a distribution confined to -1 and some arbitrary positive value instead of to -1 and $+1$, and that will be done in the next chapter, but the relative simplicity of the theory of red-and-black justifies this separate chapter. A parallel special case with $w = 1/2$ but with the possible gain different from the possible loss is also briefly treated in the present chapter.

If $w < 1/2$, the strong law of large numbers suggests that a consistent policy of small bets is a bad strategy. This may stimulate an interest in large bets. How large? It is illegal for the gambler to bet more than he possesses and plainly unwise for him to bet more than just enough to reach his goal if he wins the bet. The gambler may, then, be said to be playing boldly if he always bets his entire fortune f or enough to reach the goal, whichever is least. That is, if $f \leq 1/2$, it is bold to stake f, but if $1/2 \leq f \leq 1$, it is bold to stake $\bar{f}$. In short, it is bold to stake the minimum of f and $\bar{f}$, or, equivalently, to use the *bold gamble*

$$(1) \qquad \gamma(f) = \begin{cases} [\min (f, \bar{f})\rho + f] & \text{for } 0 \leq f \leq 1, \\ \delta(f) & \text{for } f \geq 1. \end{cases}$$

The gamble-valued function γ determines for each f a stationary strategy, called the *bold strategy at* f and designated by $\sigma(f)$. Since $\gamma(f) \in \Gamma(f)$ for all f, $\sigma(f)$ is available in the red-and-black casino Γ at f.

The main objective of this chapter is to prove that the bold strategy at f is optimal for $w < 1/2$. Incidental objectives are to establish the identity of U and a function well known in analysis and probability theory, to say something about what happens when the gambler is required to terminate at the nth move, and to treat briefly a kind of casino that might be called dual to red-and-black.

As a bonus, the study leads to new facts about a prominent family of singular distribution functions and about Bernoulli, or binomial, processes.

2. THE UTILITY OF BOLD STRATEGIES

This section is about the utility $u(\sigma(f))$ of the family of bold strategies as a function of f for $0 \leq f \leq 1$. This function will be called Q_w, when necessary, but usually simply Q. By definition then, $Q_w(f) = Q(f) = u(\sigma(f))$.

The bold strategy is almost stagnant. For both $1 - w^n$ and $1 - \bar{w}^n$ converge to 1 as n goes to infinity, and the probability of stagnation by the nth gamble is at least as large as one of these two numbers. On this account, it is easy to verify that $Q(f)$ is the (Eudoxus) probability of stagnating at 1 under $\sigma(f)$.

If $f \leq 1/2$, the gambler's first move will leave him at $2f$ with probability w and at 0 with probability $\bar{w}$. In the former case, the rest of his play is described by $\sigma(2f)$; in the latter case, play stagnates with $u = 0$. This fact and a dual one for $f \geq 1/2$ lead, through

Theorem 3.2.1, to the conclusion that

$$(1) \qquad Q(f) = \begin{cases} wQ(2f) & \text{for } f \leq 1/2, \\ w + \bar{w}Q(2f - 1) & \text{for } f \geq 1/2. \end{cases}$$

There is one and only one bounded function Q on $[0, 1]$ that satisfies (1); Q is continuous and strictly increasing; $Q(0) = 0$, and $Q(1) = 1$. These facts are easy to prove and are explicitly proved by de Rham in an interesting paper (1956–1957). Since the probability $Q(f)$ is bounded between 0 and 1, Q is characterized by (1) and has the properties cited. If $w = 1/2$, the function f obviously solves (1); so $Q_{1/2}(f) = f$ for f in $[0, 1]$.

An interesting interpretation of (1) was pointed out to us by W. Forrest Stinespring. Let $\hat{Q}(f)$ be the probability that the sequence of 0's and 1's generated by independent tosses with a *w-coin* (that is, one which produces 0 with probability w and 1 with probability $\bar{w}$) represents a number in the binary notation that does not exceed f. If f is at most 1/2, the sequence of tosses can generate a number as small as f only if the first binary digit is a 0 and the remaining digits represent a number as small as $2f$ (aside from one possibility of probability 0); if f is at least 1/2, an initial 0 ensures success, and success can also be had after an initial 1 if and only if the remaining digits represent a number as small as $2f - 1$. This all means that $\hat{Q}$ satisfies (1) and is therefore the same as Q. This interpretation shows strikingly that if $w \neq 1/2$ then Q considered as a monotonic function—more specifically, the distribution function of a probability measure—on $[0, 1]$ is singular with respect to Lebesgue measure. For, according to the strong law of large numbers, the probability measure associated with Q_w attaches probability 1 to the set of numbers in whose binary representation the relative frequency of 0's is w. Thus, as is well known, every Q_w is singular with respect to every other. In particular, if $w \neq 1/2$, then Q_w is singular with respect to $Q_{1/2}$, which corresponds to Lebesgue measure. Once it is proved that Q is U for $w < 1/2$, Q will be seen as a casino function which is as irregular, in terms of everyday intuition, as a monotonic function can be.

Interpretation of Q as a distribution function brings out a computational formula (which can of course be derived directly from (1)).

$$(2) \qquad Q((k + 1)2^{-n}) = Q(k2^{-n}) + \bar{w}^a w^b,$$

where k is a nonnegative integer less than 2^n, a is the number of 1's in the binary representation of $k2^{-n}$, and b is $n - a$. The probability argument for (2) is this: To toss a sequence representing a number between $k2^{-n}$ and $(k + 1)2^{-n}$, it is sufficient and (almost) necessary to begin with the n binary digits that represent $k2^{-n}$.

The duality

$$Q_w(f) = \bar{Q}_{\bar{w}}(\bar{f}) \tag{3}$$

can easily be established probabilistically from the definition of Q or algebraically by verifying that the right side of (3) also satisfies (1).

Several points about Q are illustrated by Table 1 for $f = k/8$, $k = 0, \cdots, 8$, and for a few other values of f, listed in binary as well as in more conventional notation. First, Q is computed by application of (1), beginning with $k = 0$ and 8 and proceeding through $k = 4$, $k = 2$ and 6, and $k = 1, 3, 5$, and 7. Next, Q is rewritten to illustrate the duality (3). Finally, the column of differences illustrates (2).

The entry for $f = 1/3$, for example, can be computed from the infinite series delivered by repeated application of (2), namely, $w^2 + w^3\bar{w} + w^4\bar{w}^2 + \cdots$. But the calculation $Q(1/3) = wQ(2/3) = w[w + \bar{w}Q(1/3)]$ is also of interest. Both types of calculation apply to express $Q_w(f)$ as a rational function of w for each rational f.

Table 1 Selected values of $Q_w(f)$

f		$Q(f)$ First form	$Q(f)$ Dual form	$\Delta Q(f)$
0	.000	0	$1 - 1$	w^3
1/8	.001	w^3	$1 - \bar{w} - w\bar{w} - w^2\bar{w}$	$\bar{w}w^2$
1/4	.010	w^2	$1 - \bar{w} - w\bar{w}$	$\bar{w}w^2$
3/8	.011	$w^2 + \bar{w}w^2$	$1 - \bar{w} - w\bar{w}^2$	$\bar{w}^2w$
1/2	.100	w	$1 - \bar{w}$	$\bar{w}w^2$
5/8	.101	$w + \bar{w}w^2$	$1 - \bar{w}^2 - w\bar{w}^2$	$\bar{w}^2w$
3/4	.110	$w + \bar{w}w$	$1 - \bar{w}^2$	$\bar{w}^2w$
7/8	.111	$w + \bar{w}w + \bar{w}^2w$	$1 - \bar{w}^3$	$\bar{w}^3$
1	1.000	1	$1 - 0$	
1/3	.0101 · · ·	$w^2(1 - w\bar{w})^{-1}$	$1 - \bar{w}(1 - \bar{w}w)^{-1}$	
2/3	.1010 · · ·	$w(1 - w\bar{w})^{-1}$	$1 - \bar{w}^2(1 - \bar{w}w)^{-1}$	
2^{-n}	.00 · · · 1	w^n	$1 - \bar{w}(1 + w + \cdots + w^{n-1})$	
$1-2^{-n}$	.11 · · · 0	$w(1 + \bar{w} + \cdots + \bar{w}^{n-1})$	$1 - \bar{w}^n$	

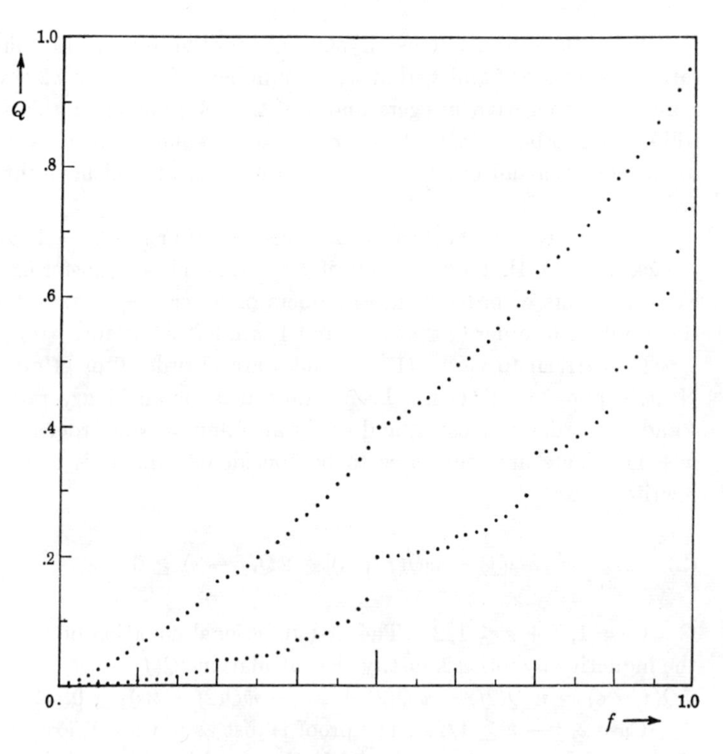

Figure 1 Values of $Q_w(f)$ for $w = .2$ and $.4$.

3. FOR $w \leq 1/2$, BOLD PLAY IS OPTIMAL.

THEOREM 1. The bold strategy at f is optimal for red-and-black casinos with $w \leq 1/2$.

Proof: In view of Theorem 2.12.1, it suffices to show that

$$Q_w(f) \geq wQ_w(f + s) + \bar{w}Q_w(f - s) \tag{1}$$

for $0 \leq f - s \leq f \leq 1$ and $w \leq 1/2$, where it is to be understood that $Q_w(f + s) = 1$ for $f + s > 1$.

The possibility that $f + s > 1$ can be set aside. For in view of the monotony of Q_w, if (1) were to fail for some $s > \bar{f}$, it would also fail for $s = \bar{f}$. Therefore, throughout the rest of the proof let it be assumed that $s \leq \bar{f}$.

Since Q_w is continuous, it suffices to establish (1) for binary rational values of f and s, that is, for numbers of the form $k2^{-n}$ with k and n nonnegative integers and $k \leq 2^n$. A number of this form will be said to be of *order at most n*. It will frequently be convenient to consider that all other numbers in the unit interval have infinite order.

At least one of the transformations $f \to 2f$ or $f \to 2f - 1$ carries f back into $[0, 1]$; both do only if $f = 1/2$. These transformations carry numbers of order n into numbers of order $n - 1$, except that the numbers of order 0, that is, 0 and 1, are left invariant.

It is trivial to verify (1) if f and s are of order 0 or of order 1. Next, suppose that (1) has been established for all binary rationals f and s of order at most n, and study an f and an s of order at most $n + 1$. There are four cases to be considered, and it is helpful to rewrite (1) as

$$Q(f) - wQ(f + s) - \bar{w}Q(f - s) \geq 0. \tag{2}$$

Case 1, $f + s \leq 1/2$. The first functional equation of (2.1) and the inductive hypothesis justify the calculation: $Q(f) - wQ(f + s) - \bar{w}Q(f - s) = w[Q(2f) - wQ(2f + 2s) - \bar{w}Q(2f - 2s)] \geq 0$.

Case 2, $f - s \geq 1/2$. The proof is just as in Case 1, except that the second functional equation of (2.1) is used instead of the first.

Case 3, $f - s < f \leq 1/2 < f + s$. Calculate thus:

$$\begin{aligned} (3) \quad Q(f) - wQ(f + s) - \bar{w}Q(f - s) & \\ &= wQ(2f) - w[w + \bar{w}Q(2f + 2s - 1)] - w\bar{w}Q(2f - 2s) \\ &= w[Q(2f) - w - \bar{w}Q(2f + 2s - 1) - \bar{w}Q(2f - 2s)]. \end{aligned}$$

Now f is necessarily greater than 1/4, for otherwise $s \leq 1/4$ and $f + s \leq 1/2$. Therefore $2f > 1/2$, and the sequence of equalities (3) can be continued thus:

$$\begin{aligned} (4) \quad &= w[w + \bar{w}Q(4f - 1) - w - \bar{w}Q(2f + 2s - 1) - \bar{w}Q(2f - 2s)] \\ &= w\bar{w}[Q(4f - 1) - Q(2f + 2s - 1) - Q(2f - 2s)] \\ &= \bar{w}[Q(2f - 1/2) - wQ(2f + 2s - 1) - wQ(2f - 2s)]. \end{aligned}$$

Since $w \le \bar{w}$, this last expression is greater than or equal to both

$$\bar{w}[Q(2f - 1/2) - wQ(2f + 2s - 1) - \bar{w}Q(2f - 2s)]$$

and

$$\bar{w}[Q(2f - 1/2) - \bar{w}Q(2f + 2s - 1) - wQ(2f - 2s)].$$

It is strictly greater than both unless $w = 1/2$ or $f - s = 0$. By the inductive assumption, the first or the second of these expressions is nonnegative, according as $2f + 2s - 1 \ge 2f - 1/2$ or $2f - 2s \ge 2f - 1/2$, that is, $s \ge 1/4$ or $s \le 1/4$.

Case 4, $f - s \le 1/2 < f < f + s$. The proof for this case is similar to that of the preceding one. ♦

Though (1) is but a very special case of the casino inequality (4.2.1), by implying that Q is the U of red-and-black casinos for $w \le 1/2$, it implies the whole casino inequality. This has certain consequences for the binomial process corresponding to independent tosses with a w-coin. For example, the special casino inequality $Q(fg) \ge Q(f)Q(g)$ implies that the conditional probability of generating a binary number less than fg, given that the number is less than g, is at least the unconditional probability that such a number is less than f.

In general, facts of this chapter and the next can be interpreted as new facts about the Bernoulli process and about the functions Q_w, which have an independent analytic interest as singular distribution functions. For literature on the Q_w see (de Rham 1956–1957).

The following instance of strict inequality in (2) will soon be used.

THEOREM 2. $Q_w(f) > wQ_w(f + s) + \bar{w}Q_w(f - s)$ if $0 < f - s < 1/2 < f + s < 1$ and $w < 1/2$.

Proof: There are two cases corresponding to Cases 3 and 4 in the proof of Theorem 1. Assume first that $f \le 1/2$. The calculation for Case 3 of Theorem 1 shows that $Q(f) - wQ(f + s) - \bar{w}Q(f - s)$ is strictly greater than a quantity which, in view of Theorem 1, is now known to be nonnegative. A dual argument applies for $f \ge 1/2$. ♦

4. OTHER OPTIMAL STRATEGIES FOR $w < 1/2$

A stake s for the fortune f is *conserving* if the lottery θ that wins s with probability w and loses s with probability $\bar{w}$ is such that $[f + \theta]$ conserves U at f, that is, if $[f + \theta]Q = Q(f)$. Thus the stake s is conserving at f if and only if

$$Q(f) = wQ(f + s) + \bar{w}Q(f - s), \tag{1}$$

which clearly requires that $f + s \leq 1$ or $f - s \geq 1$.

What are the conserving stakes available at f in $(0, 1)$, that is, the $s \leq \min(f, \bar{f})$ which satisfy (1)? The functional equations (2.1) imply that, for $f \leq 1/2$, $s = f$ is conserving, whereas, for $1/2 \leq f \leq 1$, $s = \bar{f}$ is conserving. Let $S_1(f) = f$ for $f \leq 1/2$ and $S_1(f) = \bar{f}$ for $1/2 \leq f \leq 1$, that is, $S_1(f) = \min(f, \bar{f})$. Then $S_1(f)$ is a conserving stake for the fortune f. The following observation leads to the discovery of a sequence of conserving stakes. Suppose that a gambler with a fortune $f < 1/2$ adopts a strategy σ that first maximizes the probability of reaching $1/2$ before bankrupting the gambler, and then, if he has had the good fortune to reach $1/2$, goes on to maximize the probability of reaching 1. The probability of reaching $1/2$ from f in any casino with a casino function U is at least $U(2f)$. Thus the probability of reaching 1 under σ is at least

$$U(1/2)U(2f) = Q(1/2)Q(2f) = wQ(2f) = Q(f) = U(f).$$

Therefore, σ is itself an optimal strategy.

Since red-and-black is an inclusive casino, contraction by the factor $1/2$ of an optimal strategy at $2f$ is optimal for reaching the fortune $1/2$ from f. Therefore, $S_2(f) = \frac{1}{2}S_1(2f)$ is a conserving stake for f when the space of fortunes is $[0, 1/2]$ and the goal is $1/2$. But as the preceding discussion shows, it is therefore also a conserving stake for the original red-and-black casino. Moreover, if $2f > 1/2$, $S_2(f)$ is unequal to $S_1(f)$. Thus bold strategies are not the only optimal ones.

Suppose next that a gambler with a fortune f, $1/2 \leq f \leq 1$, adopts a strategy σ that first maximizes the probability of attaining 1 before the fortune becomes as low as $1/2$, and then, if the fortune drops to $1/2$, goes on to maximize the probability of reaching 1 by adopting the bold strategy. The probability of reaching 1 under such a strategy is $U(2f - 1) + w\bar{U}(2f - 1) = Q(2f - 1) + w\bar{Q}(2f - 1) = w + \bar{w}Q(2f - 1) = Q(f) = U(f)$. Therefore, this strategy is also optimal. Again, since S_1 is a conserving-stake-valued function for red-and-black, $S_2(f) = \frac{1}{2}S_1(2f - 1)$ is conserving for $[1/2, 1]$ as the space of fortunes with 1 as the goal. Therefore, for each f in the unit interval, $S_2(f)$ is a conserving stake for the original red-and-black casino.

Define S_{n+1} by induction: $S_{n+1}(f) = \frac{1}{2}S_n(2f)$ or $\frac{1}{2}S_n(2f - 1)$ according as $f \leq 1/2$ or $f \geq 1/2$. The argument already given shows inductively that, for each n and f, $S_n(f)$ is a conserving stake.

It is easy to verify that $S_{n+1}(f)$ is the distance between f and the closest nth-order binary not necessarily different from f. Therefore $S_n(f)$ is a monotone decreasing sequence of nonnegative functions that converges to 0. For any f other than a binary rational, $S_n(f) > 0$ for all n. Therefore, for any such f there are arbitrarily small, but positive, conserving stakes, though naïve intuition about the law of large numbers might erroneously suggest that small stakes are to be avoided. It is convenient to let $S_\infty(f) = 0$; for 0 is a trivial conserving stake.

There are in fact no conserving stakes other than $S_n(f)$ and 0. Suppose otherwise. Then equality holds in (1) for some f and some positive s other than any $S_n(f)$. Choose such a pair f, s for which s is more than half as large as the supremum of all such s. Then $f - s$ and $f + s$ are strictly separated by 1/2. Why? If, for example, $f + s \leq 1/2$, an application of the first of the functional equations (2.1) implies that equality holds in (1) for $2f$ and $2s$. Furthermore, $2s$ is not $S_n(2f)$ for any n, for otherwise $s = \frac{1}{2}2s = S_{n+1}(f)$, contrary to the assumption that $s \neq S_{n+1}(f)$. A similar argument is applicable if $1/2 \leq f - s$; so $f - s < 1/2 < f + s$. Furthermore, $f - s$ is not 0, nor is $f + s$ equal to 1, for otherwise $s = S_1(f)$. But for such f and s, as Theorem 3.2 states, equality does not hold in (1). Therefore s is a conserving stake for f if and only if $s = S_n(f)$ for some n. This is recorded in the next theorem.

THEOREM 1. For $0 \leq f - s \leq f \leq 1$ and $w \leq 1/2$,

$$Q_w(f) \geq wQ_w(f + s) + \bar{w}Q_w(f - s). \tag{2}$$

Moreover, for $0 < w < 1/2$, equality holds in (2) if and only if $s = S_n(f)$ for some n (including $n = \infty$).

Now that all conserving stakes and hence all conserving gambles have been determined, it is easy to characterize all optimal strategies.

THEOREM 2. A strategy σ, available at f in $(0, 1)$ in a subfair red-and-black casino, is optimal if and only if, immediately following each of the at most denumerably many partial histories of positive σ-probability (including the vacuous one), none but conserving stakes are used, and stagnation does not occur after any such partial history except at 0 and 1.

Proof: Theorems 3.4.1 and 3.5.3 apply. The main point to check is that, if σ satisfies the hypotheses, then the probability that f_m is a binary rational converges to 1 as m approaches ∞; once f_m is a binary rational, each succeeding gamble reduces the order with certainty, until order 0 is attained. ◆

(The situation for fair red-and-black casinos is clearly quite different from that described in Theorem 2 for subfair ones.)

A stationary family of strategies $\bar{\sigma}$ is tantamount to a stake-valued function s, with $s(f) \leq f$. The characterization of optimal strategies in Theorem 2 implies that s corresponds to a stationary family of optimal strategies if and only if, for every f in $(0, 1)$, $s(f)$ is a conserving stake for f other than the trivial stake 0, and for every $f \geq 1$, $s(f) \leq f - 1$.

COROLLARY 1. A stake-valued function s determines a stationary family of optimal strategies if and only if, for every f in $(0, 1)$, there is a nonnegative integer $n = n(f)$ less than the order of f such that $s(f)$ is the distance between f and the closest nth-order binary rational (and $s(f) \leq f - 1$ for $f \geq 1$).

There clearly exist wildly irregular (for example, non-Lebesgue-measurable), conserving-stake-valued functions as well as many that are only mildly irregular. However, only the bold one (corresponding to $n(f) = 0$) is continuous, as is easily seen.

5. LIMITED PLAYING TIME

As Aryeh Dvoretzky pointed out to us, even under the requirement that play be limited to m moves, bold play is still optimal. He pointed out also that this is not true of the somewhat more general casinos to be studied in the next chapter. Here is his argument for the red-and-black casino.

Let $Q_m(f)$ be the probability of winning in m moves with the bold strategy beginning at f. The Q_m satisfy functional equations like (2.1):

$$Q_{m+1}(\tfrac{1}{2}f) = wQ_m(f),$$

$$Q_{m+1}\left(\frac{1+f}{2}\right) = w + \bar{w}Q_m(f) \tag{1}$$

for $0 \leq f \leq 1$.

The proof given for the inequality (3.1) also shows that

$$Q_{m+1}(f) \geq wQ_m(f+s) + \bar{w}Q_m(f-s) \tag{2}$$

for binary rationals f and s if $0 \leq f - s \leq f \leq 1$.

That (2) holds also when f and s are not binaries may be seen as follows. Let f_m denote for the moment the largest binary rational of order m that does not exceed f. Then

$$f_{m+1} = \begin{cases} \tfrac{1}{2}(2f)_m & \text{for } f \leq \tfrac{1}{2}, \\ \tfrac{1}{2}(1 + (2f-1)_m) & \text{for } f \geq \tfrac{1}{2}. \end{cases} \tag{3}$$

This fact and (1) imply by induction that

$$Q_m(f) = Q_m(f_m). \tag{4}$$

Finally, compute thus,

$$\begin{aligned} Q_{m+1}(f) &= Q_{m+1}(f_{m+1}) \\ &\geq Q_{m+1}(\tfrac{1}{2}(f+s)_m + \tfrac{1}{2}(f-s)_m) \\ &\geq wQ_m((f+s)_m) + \bar{w}Q_m((f-s)_m) \\ &= wQ_m(f+s) + \bar{w}Q_m(f-s). \end{aligned} \tag{5}$$

The first inequality depends on the arithmetic inequality,

$$f_{m+1} \geq \tfrac{1}{2}(f+s)_m + \tfrac{1}{2}(f-s)_m, \tag{6}$$

which is true simply because the right side of (6) is a binary of order

$m + 1$ not larger than f. Thus (2) holds without the restriction to binaries. But equality holds in (2) for the bold value of s; so

$$Q_{m+1}(f) = \sup_{s \le f} \{wQ_m(f + s) + \bar{w}Q_m(f - s)\}. \tag{7}$$

Let $U_m(f)$ be the optimal probability of winning in m moves beginning at f. Then the sequence U_m satisfies (7) in the role of Q_m. Since $U_1 = Q_1$, it follows by induction that $Q_m = U_m$ for all m, and bold strategies are seen to be optimal for limited playing time. Of course this immediately implies that $Q_w = U_\omega$. This and Theorem 4.5.1 provide a variant of the proof of Theorem 3.1 given earlier that $Q = U$.

6. THE TAXED COIN

Consider a casino in which the only game available is to bet on the outcome of a fair coin. A fixed fraction of the stake is collected as tax (or entrance fee); the remainder is paid back twice over if and only if the coin falls heads. It is convenient to denote the ratio of the possible gain to the stake by $\bar{r}/r$, where $0 < \bar{r} \le r < 1$. Thus $\Gamma(f) = \{[s\rho' + f]: s \le f\}$, where $\rho'\{-1\} = \rho'\{\bar{r}/r\} = 1/2$.

Once more the bold strategy—that for which $s(f)$ is the minimum of f and $\bar{f}r/\bar{r}$ for f in $(0, 1)$, and $s(f) = 0$ elsewhere, so that every bet has a chance to be decisive—is interesting. The utility of the bold strategy satisfies an equation closely analogous to (2.1).

$$R(f) = \begin{cases} \dfrac{1}{2} R\left(\dfrac{f}{r}\right) & \text{for } f \le r, \\[2ex] \dfrac{1}{2} + \dfrac{1}{2} R\left(\dfrac{f - r}{\bar{r}}\right) & \text{for } f \ge r, \end{cases} \tag{1}$$

which is tantamount to the pair of equations,

$$R(f) = \begin{cases} 2R(rf) & \text{for } 0 \le f \le 1, \\ 2R(r + \bar{r}f) - 1 & \text{for } 0 \le f \le 1. \end{cases} \tag{2}$$

Equation (1) has at most one bounded solution, because the discrepancy between any pair of solutions is doubled by passing from f to

f/r or to $(f - r)/\bar{r}$, whichever lies in [0, 1]. But if Q_w is the function defined by (2.1), then clearly $Q_{\bar{r}}^{-1}$ satisfies (1) so $R_r(f) = Q_{\bar{r}}^{-1}(f)$.

The fundamental inequality for Q_r with $r > 1/2$ is obtained by applying the duality (2.3) to the inequality (3.1).

$$Q_r(f) \leq rQ_r(f + s) + \bar{r}Q_r(f - s) \tag{3}$$

for $0 \leq f - s \leq f + s \leq 1$ and $1 \geq r \geq 1/2$.

This is equivalent to

$$R_r(g) \geq \tfrac{1}{2}R_r(f + t\bar{r}) + \tfrac{1}{2}R_r(g - tr) \tag{4}$$

for $0 \leq g - tr \leq g \leq 1$.

Therefore the bold strategy is optimal for this casino, and R is its casino function. It is also easy to describe all optimal strategies and to show that bold strategies are optimal even for limited playing time.

To do the latter, prove that

$$R_m(f) = R(f)_m, \tag{5}$$

where $R_m(f)$ is the probability of success in m moves with the bold strategy beginning at f, and $R(f)_m$ is the largest binary rational of order m that does not exceed $R(f)$. Since any further details would bore you and us, we move on to the next chapter and a more general type of casino.

6
PRIMITIVE CASINOS

1. INTRODUCTION. This chapter is about a significant generalization of the red-and-black casinos, which we shall call primitive casinos because information about them and their casino functions has important implications for all casinos and casino functions (Section 7).

The wording and organization of this chapter take advantage of many parallels with Chapter 5, which this chapter generalizes. It is possible to cover swiftly here much that will be easy in the light of Chapter 5 and also to give new insights into that chapter.

In a *primitive casino*, once again, all the gambler can do is to stake an amount s, losing it with probability $\bar{w}$ and winning with probability w, $0 < w < 1$. The difference is that now his fortune is not necessarily increased by s, when he wins a bet, but by some fixed multiple of s, which it is convenient to write $\bar{r}s/r$. Formally, if ρ is now the measure that attaches w to $\bar{r}/r$ and $\bar{w}$ to -1 for some r in $(0, 1)$, then $\Gamma(f) = \{[s\rho + f]: 0 \leq s \leq f\}$. The possibilities $w = 0$ or 1, or $r = 0$ or 1, would not be interesting; $r = 1/2$ is red-and-black; $w = 1/2$ is the taxed coin, also treated in Chapter 5. A primitive casino is easily seen to be superfair, fair, or subfair according as $w > r$, $w = r$, or $w < r$; this is one advantage of the notation in terms of r.

To play boldly at f in $(0, 1)$ is, as before, to stake as much as possible without risk of overshooting the goal, that is, to make the largest bets that are both legal and reasonable. The *bold gamble at* f in the primitive casino Γ is

$$(1) \qquad \gamma(f) = \begin{cases} \left[\min\left(f, \frac{rf}{\bar{r}}\right)\rho + f\right] & \text{for } 0 \le f \le 1, \\ \delta(f) & \text{for } f > 1, \end{cases}$$

and the stationary family of strategies σ determined by γ are the *bold strategies*.

The main objective of this chapter is to prove that $\sigma(f)$ is optimal for $w < r$. Other objectives, in analogy with Chapter 5, are to determine $U(f)$, to find all optimal strategies, and to say something surprising about what happens when the gambler is required to terminate at the nth move. There are still other objectives and by-products, without analogy in Chapter 5, that it would not be practical to indicate in advance. More new inequalities for binomial processes are a bonus.

2. THE UTILITY OF BOLD STRATEGIES

Let $S_{w,r}$, or S for short, be the utility of the bold strategies as a function of f.

The derivation of (5.2.1) is easily generalized to show that

$$(1) \qquad S(f) = \begin{cases} wS\left(\frac{f}{r}\right) & \text{for } f \le r, \\ w + \bar{w}S\left(\frac{f-r}{\bar{r}}\right) & \text{for } f \ge r. \end{cases}$$

This is obviously equivalent to the more handsome equations,

$$(2) \qquad \begin{aligned} \frac{S(f)}{w} &= S\left(\frac{f}{r}\right) && \text{for } f \le r, \\ \frac{S(f) - w}{\bar{w}} &= S\left(\frac{f-r}{\bar{r}}\right) && \text{for } f \ge r, \end{aligned}$$

and to the useful

$$(3) \qquad S(rf) = wS(f) \qquad S(r + \bar{r}f) = w + \bar{w}S(f) \qquad \text{for } 0 \le f \le 1.$$

These pairs of equations are as obviously equivalent to the duality-revealing pair,

$$wS(f) = S(rf) \qquad \bar{w}\bar{S}(\bar{f}) = \bar{S}(\overline{\bar{r}\bar{f}}). \tag{4}$$

One and only one bounded function of f on $[0, 1]$ satisfies (1). This solution is continuous and strictly increasing; $S(0) = 0$, and $S(1) = 1$. These facts could apparently be proved by extending the method of de Rham (1956–1957), but it is more convenient here to infer them mainly from the case treated by him and already used in Chapter 5.

The uniqueness is easy to prove directly, as it was for the taxed coin. For the difference ΔS between two solutions of (1) satisfies

$$\Delta S\left(\frac{f}{r}\right) = \frac{\Delta S(f)}{w} \qquad \text{for } f \leq r, \tag{5}$$

$$\Delta S\left(\frac{f - r}{\bar{r}}\right) = \frac{\Delta S(f)}{\bar{w}} \qquad \text{for } f \geq r, \tag{6}$$

so that ΔS must be identically 0 or unbounded.

In the notation of Chapter 5, $Q_w(Q_r^{-1}(f))$ is easily seen to satisfy (1); so

$$S_{w,r}(f) = Q_w(Q_r^{-1}(f)). \tag{7}$$

Therefore S does have the continuity and monotony anticipated.

The construction (7) will be useful later, especially when expressed as a pair of parametric equations,

$$S(a) = Q_w(a) \qquad f(a) = Q_r(a) \qquad \text{for } 0 \leq a \leq 1. \tag{8}$$

In view of (8), the interpretation of $Q_w(f)$ as the probability of tossing a number as small as f with a w-coin has a counterpart here, but a clumsy one. Namely, $S_{w,r}(f)$ is the probability that a w-coin will produce a number as small as that number a for which an r-coin produces numbers as small as a with probability f.

It is easy to verify these relations among the $S_{w,r}$:

$$S_{w,r}(f) = S_{r,w}^{-1}(f) = \bar{S}_{\bar{w},\bar{r}}(\bar{f}) = \bar{S}_{\bar{r},\bar{w}}^{-1}(\bar{f}). \tag{9}$$

In the language of Section 4.6, $S_{w,r}$ and $S_{\bar{r},\bar{w}}$ are a pair of dual functions. Unless $w = r$, in which case $S(f) = f$, $S_{w,r}$ is singular. To see this, note that, according to the strong law of large numbers, Q_w and Q_r are mutually singular, and then apply the following simple theorem.

THEOREM 1. If Q and Q^* are continuous distribution functions, and Q is strictly increasing, then Q and Q^* are mutually singular if and only if Q^*Q^{-1} is singular with respect to Lebesgue measure.

Proof: The following four assertions are plainly equivalent.

(*a*) Q and Q^* are mutually singular.

(*b*) There is a set A for which $Q^*(A)$ has Lebesgue measure 0 and $Q(A)$ has Lebesgue measure 1.

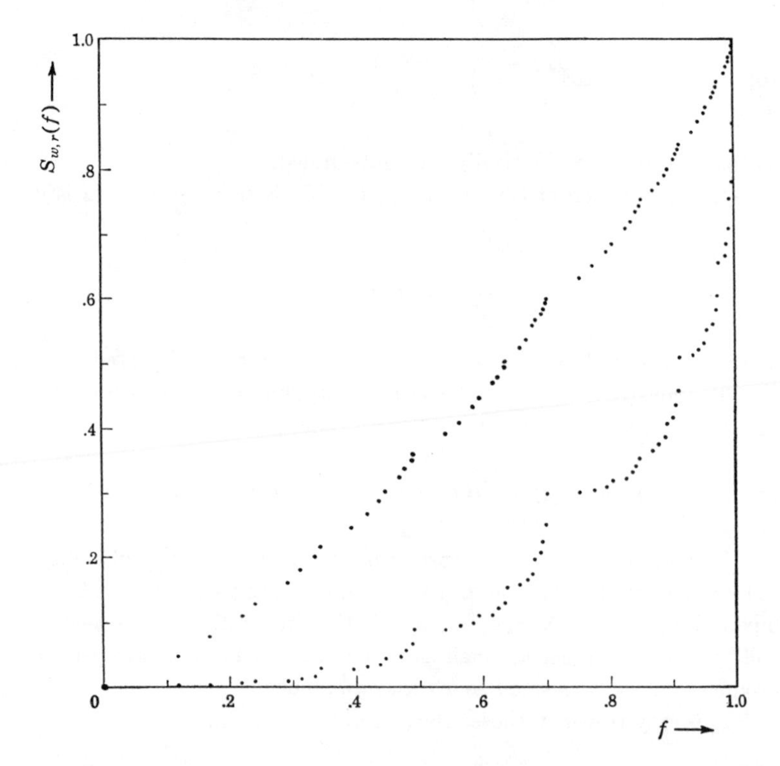

Figure 1 Values of $S_{.3,\,.7}(f)$ and $S_{.6,\,.7}(f)$.

(*c*) There is a set B of Lebesgue measure 0 for which $Q^*Q^{-1}(B)$ has Lebesgue measure 1.

(*d*) Q^*Q^{-1} is singular with respect to Lebesgue measure. ♦

Except in the trivial cases, $w = w'$ and $r = r'$ or $w = r$ and $w' = r'$, we suspect that $S_{w,r}$ is singular with respect to $S_{w',r'}$. Theorem 1 makes this clear for $r = r'$. And Theorem 8.2 says that the functions are not identical except in the trivial cases.

3. BOLD PLAY IS OPTIMAL FOR SUBFAIR PRIMITIVE CASINOS.

Parallel with Section 5.3, it is important to prove that

$$T(f, g) = S(rf + \bar{r}g) - wS(f) - \bar{w}S(g) \geq 0 \tag{1}$$

for $f \geq g$ and (as always in this section) $w \leq r$. This will show that $S = U$, that bold strategies are optimal, and that S satisfies the whole casino inequality. The case $w = r$, when S is fair, is trivial as before.

Our only proof of (1) is much more difficult and roundabout than that of the analogous (5.3.1).

The problem will be transformed several times before (1) is finally proved. First of all, attention is shifted from T to a new function T'.

$$T'(f, g) = \begin{cases} S(f + g) - S(f) - S(g) & \text{for } f + g \leq 1, \\ S(f + g - 1) - S(f) - S(g) + 1 & \text{for } f + g \geq 1. \end{cases} \tag{2}$$

If $T(f, g)$ were ever negative with $f \geq g$, then, as will be shown, $T'(f', g')$ would also be negative for some f', g' in $[0, 1]$. (The converse is also interesting to know and easy to prove but of no logical importance for the proof of (1).) The program now is: (*a*) to confirm that violation of (1) implies a negative value of T'; and (*b*) to show that negative values of T' are impossible, or, equivalently, that S is superadditive and $\bar{S}(\bar{f})$ is subadditive in f.

If there is any solution of $T(f, g) < 0$ with $f \geq g$, then there is one with $f > r > g$. Consider, in fact, any solution with $f - g$

nearly as large as possible. Since $T(f, f) = 0$, $f - g > 0$. If $f \leq r$ or $g \geq r$, then application of (2.1) leads to a new solution substantially more widely separated, which is impossible.

Suppose then that (1) is violated by f, g with $f > r > g$, and calculate thus:

$$wS(f) = w^2 + w\bar{w}S\left(\frac{f - r}{\bar{r}}\right); \tag{3}$$

$$\bar{w}S(g) = w\bar{w}S\left(\frac{g}{r}\right). \tag{4}$$

$$\begin{aligned} S(rf + \bar{r}g) &= wS\left(\frac{rf + \bar{r}g}{r}\right) \\ &= w^2 + w\bar{w}S\left(\frac{f + \frac{\bar{r}}{r}g - r}{\bar{r}}\right) \\ &= w^2 + w\bar{w}S\left(\frac{f - r}{\bar{r}} + \frac{g}{r}\right) \end{aligned} \Bigg\} \text{ for } rf + \bar{r}g \leq r, \tag{5}$$

$$\begin{aligned} &= w + \bar{w}S\left(\frac{rf + \bar{r}g - r}{\bar{r}}\right) \\ &= w + w\bar{w}S\left(\frac{f - r}{\bar{r}} + \frac{g}{r} - 1\right) \end{aligned} \Bigg\} \text{ for } rf + \bar{r}g \geq r.$$

According to (3), (4), and (5), if $f > r > g$ and $T(f, g) < 0$, then T' is negative at the pair of arguments $(f - r)/\bar{r}$ and g/r.

The problem now is to show that $T'(f, g) \geq 0$ for all f, g in $[0, 1]$. This will entail three or more reformulations of the problem, depending on how you count them.

First, what does it mean in terms of Q_w and Q_r to say that $T' \geq 0$? If, for example, $f + g \leq 1$, let a, b, c be such that $Q_r(a) = f + g$, $Q_r(b) = f$, $Q_r(c) = g$. Then, introducing a new function D,

$$D(r) = Q_r(a) - Q_r(b) - Q_r(c) = 0, \tag{6}$$

and

$$T'(f, g) = D(w) = Q_w(a) - Q_w(b) - Q_w(c) \geq 0. \tag{7}$$

So it has to be proved that no negative value of $D(w) = D(w; a, b, c)$ precedes any zero of $D(w)$ for any a, b, c in $[0, 1]$.

The case $f + g \geq 1$, on the other hand, requires it to be proved that $D(w) + 1$ also has no negative values prior to any of its zeros. But since $Q_w(a) = 1 - Q_{\bar{w}}(\bar{a})$, according to (5.2.3) or the more general (2.9), $D(w; a, b, c) + 1 = -D(\bar{w}; \bar{a}, \bar{b}, \bar{c})$. Thus the condition on $D(w) + 1$ amounts to requiring that no positive value of $D(w)$ succeeds any of its zeros.

All in all, what must be shown is that, for w in $(0, 1)$, all the zeros of $D(w)$ are preceded by only nonnegative, and succeeded by only nonpositive, values.

(Though the fact does not enter logically into the present argument, it may be clarifying to point out here that Q, and therefore also D, is analytic in w; so D must actually have at most one zero in $(0, 1)$ if it is not identically 0. Indeed, in view of (5.2.2), $Q_w(a)$ is expressible as a series, terminating or infinite, of the form $\Sigma\, \bar{w}^n w^{a(n)}$, where $a(n)$ is a nondecreasing sequence of positive integers. The terms of the series are dominated by α^n in the complex domain

$$\{w\colon |w| < \alpha, |1 - w| < \alpha\},$$

so the series converges uniformly to an analytic function in each such domain with $\alpha < 1$. Incidentally, $Q_w(1/3)$ is singular just at the two complex numbers for which $|w| = |1 - w| = 1$, according to Table 5.2.1.)

It will be enough to show merely for binary rationals a, b, c that, as a function of w, D has no positive value after, or negative value before, any of its zeros. To see that this will suffice, suppose, for example, that, for some real numbers a_1, b_1, c_1, the function D_1 had a negative value before one of its zeros, where

$$D_1(w) = D(w; a_1, b_1, c_1).$$

Thus, for some w_1 and w_2 with $w_1 < w_2$, $D_1(w_1) < 0 = D_1(w_2)$. Choose binary rationals a, b, c, for which $D(w_1) < 0$ and $D(w_2) \geq 0$. For example, choose a, b, c, close, respectively, to a_1, b_1, c_1 with $a \geq a_1$, $b \leq b_1$, and $c \leq c_1$. Then, for some w_3 with $w_1 < w_3 \leq w_2$, $D(w_3) = 0$. Hence a negative value of D would precede one of its zeros. It thus suffices to restrict attention to binary rationals, as asserted. This program for binary rationals involves a digression about $Q_w(a)$ for binary rational a, that is, a of the form $x2^{-n}$ with x and n nonnegative integers and $x \leq 2^n$.

If x is a nonnegative integer, let $k(x)$ be the number of 1's in the binary expansion of x; equivalently, $k(0) = 0$, $k(2x) = k(x)$, and $k(2x + 1) = k(x) + 1$.

Let $N(j; x)$ be, for all integers j (including negative ones), the number of nonnegative integers y for which $y < x$ and $k(y) = j$. Equivalently, $N(j; 0) = 0$ for all j, and

$$N(j; x + 1) = N(j; x) + \delta(j, k(x)), \tag{8}$$

where $\delta = 1$ or 0 according as $k(x)$ is or is not j.

The relevance of N is that if $a = x2^{-n}$ then

$$Q_w(a) = w^n \sum_j N(j; x) \left(\frac{\bar{w}}{w}\right)^j, \tag{9}$$

a relation easily proved by means of (5.2.2). (Incidentally, N is easily seen to be characterized by (9).)

In view of the definition of $N(j; x)$,

$$N(j; 2x) = N(j; x) + N(j - 1; x); \tag{10}$$

$$N(j; 2x + 1) = N(j; x + 1) + N(j - 1; x). \tag{11}$$

Return now to the study of $D(w; a, b, c)$ with a, b, c binary rationals but with the new notations: $a = x2^{-n}$, $b = y2^{-n}$, $c = z2^{-n}$, $v = \bar{w}/w > 0$.

$$\begin{aligned} D'(v; x, y, z) &= w^{-n} D(w; a, b, c) \\ &= \sum_j [N(j; x) - N(j; y) - N(j; z)]v^j \\ &= \sum_j M(j; x, y, z)v^j. \end{aligned} \tag{12}$$

In these terms, the problem is to show that the polynomial $D'(v)$ is never positive before, or negative after, a zero, where it is to be understood that $D'(v)$ is considered meaningful only for strictly positive v. More will actually be shown, namely, that the coefficients $M(j)$ of D' are *Cartesian*, that is, if $M(j) < 0 < M(j')$, then $j < j'$. This is more than is required; for if M is Cartesian, then D' either has the same sign for all positive v, or is identically 0, or has a single (positive) zero preceded by negative and followed by positive values, according to "Descartes' rule of signs". (See, for example, (Dickson 1939, p. 77) or, for historical sidelights (Descartes 1925, p. 160).)

The proof that M is Cartesian will be given by induction on x, y, z according to the following two-step program:

Step 1. M is Cartesian for $x, y, z \leq 2 = 2^0 + 1$.

Step 2. If M is Cartesian for all eight triples of the form $x + \xi$, $y + \eta$, $z + \zeta$, where $\xi, \eta, \zeta = 0$ or 1, then M is also Cartesian for the eight triples $2x + \xi$, $2y + \eta$, $2z + \zeta$.

Proof of these two steps will effect the induction, thus: According to Step 2, if M is Cartesian for all $x, y, z \leq 2^n + 1$, then M is also Cartesian for $x, y, z \leq 2^{n+1} + 1$; and Step 1 starts the induction with $n = 0$.

Proof of Step 1: The following general facts go a long way toward proving that M is Cartesian in the 27 instances of $0 \leq x, y, z \leq 2$. $M(x, y, z) = M(x, z, y)$. $M(x, y, 0) = M(x, 0, y)$ and never changes sign with changes in j, because $N(j; x)$ is nondecreasing in x. $M(x, y, z)$ is nonpositive if $x \leq y$ or $x \leq z$, for the same reason.

There remains only to study the instances of $2 \geq x > y \geq z > 0$, and of these there exists but one, $x = 2$ and $y = z = 1$. But $M(j; 2, 1, 1) = N(j; 2) - 2N(j; 1)$, which is 0 for $j < 0$, $1 - 2 = -1$ for $j = 0$, $1 - 0 = 1$ for $j = 1$, and is $0 - 0 = 0$ for $j > 1$; so this sequence is Cartesian, and Step 1 is proved.

Proof of Step 2: Introducing abbreviations and exploiting (10) and (11), what has to be proved here is that $M + \delta + M^*$ is Cartesian if all eight instances of $M + \delta$ are Cartesian, where $M = M(j) = M(j; x, y, z)$, $\delta = \delta(j) = \xi\delta(j, k(x)) - \eta\delta(j, k(y)) - \zeta\delta(j, k(z))$, and $M^* = M^*(j) = M(j - 1)$.

Let j be called a *plus*, a *minus*, or a *zero* of a sequence according as the sequence is positive, negative, or 0 at j. To illustrate, a sequence is Cartesian if and only if none of its minuses exceeds any of its pluses.

If all eight instances of $M + \delta$ are Cartesian, then, as will be shown, no minus of $M + \delta$ exceeds any plus of M, and every plus of

$M + \delta$ exceeds every minus of M. (Note the slight asymmetry of the two parts of this statement. Neither symmetric form is of interest to us; one of them is too weak, and the other is false.)

First, suppose that $M + \delta$ is negative at j and M is positive at j' with $j > j'$; then $k(y)$ or $k(z)$ or both must be j. Consider the instance of $M + \delta$ that arises with $\xi = 0$, and η and $\zeta = 1$ or 0 according as the corresponding $k(y)$ or $k(z)$ is or is not j. This instance of $M + \delta$ is identical with M except at j, where it is negative. It therefore is positive at j' and negative at j; so it is not Cartesian, contrary to hypothesis. By practically the same argument, a plus of $M + \delta$ cannot be exceeded by a minus of M either, nor can it ever equal a minus of M, since $M(j)$ is an integer and $\delta(j) \leq 1$.

Obviously M^* resembles $M + \delta$ in that no minus of M^* exceeds any plus of M and every plus of M^* exceeds every minus of M, because M is Cartesian. Therefore $M + \delta + M^*$ has this same property.

Now suppose that $M + \delta + M^*$ is not Cartesian, but is negative at j and positive at j' with $j > j'$. In view of the preceding paragraph, j exceeds no plus of M and j' exceeds every minus of M. Since $j' < j$, j' is exceeded by every plus of M, and $M(j' - 1) = M^*(j') \leq 0$; so $M + \delta$ is positive at j'. Since $j' < j$, j exceeds every minus of M by more than 1, and $M(j - 1) = M^*(j) \geq 0$; so $M + \delta$ is negative at j. This, however, contradicts the hypothesis that $M + \delta$ is Cartesian.

The induction is complete; $M(x, y, z)$ is Cartesian in j for all x, y, z; and the aims of the section are attained.

The conclusion that bold play is optimal for a casino generated by a lottery ρ plainly applies also to a casino generated by a mixture of ρ with the trivial lottery $\delta(0)$. Possibly it applies to any subfair casino generated by a mixture of two lotteries, one of which has no chance for positive gain and the other of which is $\delta(x)$ for some positive x. Nevertheless, boldly we conjecture that essentially the only one-game subfair casinos for which bold play is optimal are the primitive casinos.

4. DIGRESSION ABOUT N

Since the study of primitive casinos has led to the function N and a curious inequality concerning it, some may be interested in studying N and the inequality a little further with us, although we ourselves have done scarcely more than raise some natural questions. This section can freely be postponed or omitted.

Since the fact that $N(a) - N(b) - N(c)$ is Cartesian seems to be just about needed to prove the inequality (3.1), and since that inequality actually implies the much more general casino inequality (4.2.1), maybe the full casino inequality corresponds to the Cartesianness of a sequence more general than M. Specifically, the full casino inequality says that, for $b \geq c$,

$$C(w) = Q_w(a) - Q_w(d)Q_w(b) - \bar{Q}_w(d)Q_w(c) \tag{1}$$

has no negative values prior to a zero; so maybe the coefficients of the polynomial $(1 + v)^{n+m}C((1 + v)^{-1})$, for each a, b, c of order m and d of order n, are Cartesian. We do not pursue the question now, but mention it in case N proves to be mathematically interesting apart from gambles.

Again, it is natural to suspect that all differences of the form

$$\sum_{r=1}^{m} N(j; x_r) - \sum_{s=1}^{n} N(j; y_s) \tag{2}$$

are Cartesian-like in having at most a limited number, say $\lambda(m, n)$, of changes in sign. We suspect that $\lambda(m, n)$ is min (m, n) if $m \neq n$ and is $m - 1$ if $m = n$. Examples attaining these bounds are easily constructed on the basis of $N(j; 1) = \delta(j, 0)$, $N(j; 2^n) - N(j; 2^n - 1) = \delta(j, n)$. The bounds have already been shown, more or less explicitly, to be correct for $m = 0$ and 1; for the proof that M is Cartesian does not really use the fact that $n = 2$.

The function $N(j; x)$ is conveniently expressible in terms of binomial coefficients and the binary expansion for x,

$$x = \sum_{r=0}^{k(x)-1} 2^{p(r;\, x)}, \tag{3}$$

where $p(r; x)$ is strictly decreasing in r. Equation (3), of course, defines a one-to-one correspondence between positive integers x and strictly decreasing sequences of nonnegative integers p. According to the definition of N or according to (3.10) and (3.11),

$$N(j; x) = \sum_r \binom{p(r; x)}{j - r} \tag{4}$$
$$= N'(j; p(x)),$$

where an established notation for binomial coefficients and an evident abbreviation have been introduced. Thus the Cartesianness of $N(a) - N(b) - N(c)$ is seen as an inequality pertaining to certain sums of binomial coefficients. The following special case is interesting in that it relates two, not three, sums.

$$\sum_r \binom{p(r)}{j - r} - \sum_s \binom{q(s)}{j - s + k} \tag{5}$$

is Cartesian in j for each $k \geq 0$, for (5) can be seen as an instance of $M(j + k)$ by adding

$$\sum_{t=1}^{k} \binom{p(0) + t}{j + t} \tag{6}$$

to (5) and then subtracting it.

Of course, $N(x)$ is never the same function of j for two different x; indeed, $\Sigma\, N(j; x) = x$. Therefore $N'(p)$ is never the same function of j for two different sequences p. For a fixed j, however, it is quite possible to have the same value of N (or N') for many x (or p). For each j except $j = 0$, $N(j; x)$ is nondecreasing and assumes all nonnegative integers as values. It has the value 0 from $x = 0$ through $x = 2^j - 1$. For $n > 0$, $N(j; x) = n$ from just beyond the nth x for which $k(x) = j$ through the $(n + 1)$st. This latter is also characterized as the only solution x of $N(j; x) = n$ for which $k(x) = j$. Clearly now, for each positive j and nonnegative n, there is a unique sequence of nonnegative integers $p(0) > p(1) > \cdots > p(j - 1)$ such that

$$n = \binom{p(0)}{j} + \binom{p(1)}{j - 1} + \cdots + \binom{p(j - 1)}{1}. \tag{7}$$

Kruskal (1963) and Lehmer (1964) also noticed this manner of "writing" n in terms of $p(0), \cdots, p(j-1)$ and use it in certain combinatorial problems.

Certain other studies of combinatorial problems in the literature, for example (Riley and Hatcher 1958, 1959), bear some, though perhaps only superficial, resemblance to our study.

5. OTHER OPTIMAL STRATEGIES FOR $w < r$

This section determines all optimal strategies for subfair primitive casinos.

THEOREM 1. The S of a subfair primitive casino satisfies

$$S(f+g) \geq S(f) + S(g) \tag{1}$$

and

$$\bar{S}(\bar{f}+\bar{g}) \leq \bar{S}(\bar{f}) + \bar{S}(\bar{g}) \tag{2}$$

for $0 \leq f \leq f+g \leq 1$. The inequalities are strict except in the trivial cases $f = 0$, $g = 0$, $w = 0$, or $r = 1$.

Proof: The superadditivity of S and the subadditivity of $\bar{S}(\bar{f})$ are mere paraphrases of the nonnegativity of $T'(f, g)$ in Section 3. If equality holds in (1), then equality holds in (3.6) and in (3.7). Since $w < r$, D is identically 0 in $(0, 1)$ according to the parenthetical remark below (3.7), and $D(1)$ is 0 because of the easily verified continuity of D in $[0, 1]$. Therefore either b or c is 0, and so f or g is 0. A parallel argument handles equality in (2). ♦

COROLLARY 1. $S(rf + \bar{r}g) > wS(f) + \bar{w}S(g)$ for $0 < g < r < f < 1$ and $0 < w < r < 1$.

As f ranges over the binary rationals of order n, $Q_r(f)$ ranges over the *r-ary numbers of order n*. Any two successive r-ary numbers of order n are the endpoints of one of the 2^n *r-ary intervals of order n*.

The next result, which in effect characterizes all conserving stakes for subfair primitive casinos, is easy to establish by arguments parallel to those for Theorem 5.2.

THEOREM 2. $S(rf + \bar{r}g) \geq wS(f) + \bar{w}S(g)$ for $f \geq g$ and $0 \leq w \leq r \leq 1$. For $1 \geq f > g$ and $0 < w < r < 1$, the inequality is strict unless f or g is an endpoint of an r-ary interval that contains $rf + \bar{r}g$.

Now that all conserving stakes have been determined, it is easy to characterize all optimal strategies. Indeed, Theorem 5.4.2 holds for subfair primitive casinos, and a parallel to Corollary 5.4.1 characterizes all stationary families of optimal strategies.

6. LIMITED PLAYING TIME

Aryeh Dvoretzky discovered and showed us that truncating play at the mth move reveals a difference between the general primitive casino and the special types studied in the last chapter: bold strategy is no longer necessarily optimal for limited playing time. We do not attempt a systematic exploration of the situation but simply present two examples by Dvoretzky, with his permission, and point out as well as we can where the demonstration of the efficacy of bold strategies for red-and-black and for the taxed coin does not generalize to all primitive casinos.

Example 1. Suppose $w < r < 1/2$, and consider what can be done in three moves beginning at any f in the half-open interval $[r - r\bar{r}(\bar{r} - r), r)$. The bold strategy wins in three moves if and only if the first move and at least one of the next two moves are successful. The probability of attaining the goal in three moves under the bold strategy is therefore $w(w + w\bar{w})$.

But a better strategy is to stake only enough on the first move to reach the fortune $r + r\bar{r}$ if successful, and to play the remaining two moves boldly. This modification does no harm; the gambler still attains the goal in all those cases in which he attained it under the bold strategy. In addition, even if the gambler's first move is unsuccessful, he will now find himself with $f - [(r + r\bar{r}) - f]r/\bar{r} = [f - r^2(1 + \bar{r})]/\bar{r} \geq r^2$, a fortune from which he will attain the goal if his two remaining bold moves are successful. This modification therefore adds $\bar{w}w^2$ to the probability of attaining the goal.

Example 2. This example with $r > 1/2$ is more delicate. Suppose

$2/3 < w < r < 2^{-1/2}$, and consider what can be done in four moves beginning at $f = r^2 + 2\bar{r}r^3$. Notice that $r - f = r\bar{r}(1 - 2r^2)$ and $r^2 < 1/2$. The probability of reaching the goal in four bold moves is therefore $w[w + \bar{w}(w + \bar{w}w)] = w^2[1 + \bar{w}(1 + \bar{w})]$.

But now suppose that the gambler's first move is to stake $s = r^2 + \bar{r}r^3 - r^4$ and the remaining three moves are bold. If the first move is unsuccessful, the gambler finds himself with $f - s = r^3$; and if successful, with $f + \bar{r}s/r = r + \bar{r}r^2$. The probabilities of attaining the goal from these two positions in the three remaining moves are easily seen to be w^3 and $w + \bar{w}w^2$. The total probability of attaining the goal under the modified policy is therefore $\bar{w}w^3 + w(w + \bar{w}w^2) = w^2[1 + 2\bar{w}w] =$ $w^2[1 + \bar{w}(1 + \bar{w})] + w^2\bar{w}[3w - 2]$, and the modified strategy is better than the bold strategy, since $w > 2/3$.

According to arguments in Sections 5.5 and 5.6, bold strategies are optimal even for limited playing time if $r = 1/2$ (red-and-black) or if $w = 1/2$ (the taxed coin). Up to a point, these arguments extend to the general primitive casino, although, as Examples 1 and 2 show, there are limits to this extension. Specifically, the inequality (5.5.6) is applied to fortunes in the arguments for $r = 1/2$ and to the probabilities $S_{1/2,r} = R_r$ in the argument for $w = 1/2$. We think that these applications of (5.5.6) must be viewed as two special devices that are without counterpart in the general situation.

7. PRIMITIVE CASINOS IN RELATION TO GENERAL CASINOS

A large class of casinos is generated by forming unions of primitive casinos, that is, by allowing the gambler at each move to choose w and r from among a set of possibilities P and then to choose a stake s (with $s \leq f$) to be bet in the primitive casino with parameters w and r. In self-explanatory notation, such a union of primitive casinos is determined by

$$\Gamma_P(f) = \bigcup_{(w,r)\in P} \Gamma_{w,r}(f). \tag{1}$$

For greater harmony in this section, let the domain of w and r be extended to the values 0 and 1 that have heretofore been excluded as trivial. The new possibilities with $r = 0$ do not quite fall within the definition of Section 1, but they should clearly refer to those casinos in which each gamble results in the status quo f with probability $\bar{w}$ and in any positive gain of the gambler's choice with probability w.

According to the sufficiency proof of Theorem 4.2.1, if U is a casino function, it is the U of Γ_P, where P consists of the pairs $(U(r), r)$ for all r in $[0, 1]$. This remark, along with the corollary below, somewhat justifies our adoption of the word "primitive".

THEOREM 1. If U is a casino function (or even a bounded, nonzero solution of (4.3.1) and (4.3.2)) and $w \le U(r)$, then $S_{w,r}(f) \le U(f)$ for all f. To put it a little differently, if $U(r) \ge S_{w,r}(r)$, then U majorizes $S_{w,r}$.

Proof: Equation (2.3), and the following instances of the special casino inequalities, (4.3.1) and (4.3.3), show that if $U(g) \ge S_{w,r}(g)$ for a given g, the same is true for rg and for $r + \bar{r}g$.

$$U(rg) \ge U(r)U(g) \ge wU(g) \ge wS_{w,r}(g) = S_{w,r}(rg); \tag{2}$$

$$\begin{aligned} U(r + \bar{r}g) &\ge U(r) + \bar{U}(r)U(g) \\ &\ge w + \bar{w}U(g) \\ &\ge w + \bar{w}S_{w,r}(g) \\ &= S_{w,r}(r + \bar{r}g). \end{aligned} \tag{3}$$

This settles the proof when U is continuous; the two discontinuous cases are trivial. ◆

The theorem can be recast to say that primitive casino functions $S_{w,r}$ are primitive for casino functions in somewhat the same way that linear functions are primitive for convex functions, thus:

COROLLARY 1. Every casino function U is at each f the maximum of those primitive casino functions S that U majorizes throughout $[0, 1]$.

The converse to Corollary 1 does not hold. For consider the

maximum of a casino function and a primitive casino function that crosses it. This example shows a bit more: the maximum of two casino functions need not be a casino function.

COROLLARY 2. If neither of the primitive casino functions $S_{w,r}$ and $S_{w',r'}$ majorizes the other, then $S_{w,r}(r') < w'$ and $S_{w',r'}(r) < w$. Consequently, for some f between r and r', $S_{w,r}(f) = S_{w',r'}(f)$.

THEOREM 2. Let U be the strictly increasing casino function of a casino Γ, and suppose that U majorizes the primitive casino function $S_{w,r}$. Then U is excessive for the primitive casino $\Gamma_{w,r}$; equivalently, U is the casino function of the union of Γ with $\Gamma_{w,r}$. If $U(r) > w$, then the only gamble in $\Gamma_{w,r}(f)$ conserving for the union at f is $\delta(f)$.

Proof: Let $\gamma \in \Gamma_{w,r}(f)$. Then

$$\begin{aligned}\gamma U &= wU(f + s\bar{r}) + \bar{w}U(f - sr)\\ &= S_{w,r}(r)U(f + s\bar{r}) + \bar{S}_{w,r}(r)U(f - sr)\\ &\leq U(r)U(f + s\bar{r}) + \bar{U}(r)U(f - sr)\\ &\leq U(f),\end{aligned}$$

where the last inequality is an application of the casino inequality (4.2.1). Therefore U is excessive for $\Gamma_{w,r}$. The rest of the proof is obvious, especially in the light of Corollary 2.12.1. ◆

The U of the union of two casinos Γ_1 and Γ_2 such that U_1 majorizes U_2 can sometimes exceed U_1 (Theorem 9.2.5). Therefore not all casinos can replace the primitive casinos in Theorem 2; perhaps none other than unions of primitive casinos can. Of course in the special case that U is fair, if Γ' is any casino whose U' is majorized by U, then Γ' and U' can replace $\Gamma_{w,r}$ and $S_{w,r}$ in Theorem 2. Perhaps no subfair U has this property of the fair U.

8. MORE ABOUT PRIMITIVE CASINO FUNCTIONS

The functions $S_{w,r}$ solve one of the simplest types of gambler's problems; they are determined by simple functional equations; they satisfy curious functional inequalities; they stand in a special relation to casino functions in general; and, unlike the "special functions" usually evoked by simple problems, they are in a sense highly irregular. For these reasons, study of the $S_{w,r}$, alone, and with respect to one another, seems promising. This section digresses to report and apply some facts on this topic.

First, to review: For w and r in $(0, 1)$, $S_{w,r}$ is characterized as the unique bounded solution of (2.1) for f in $[0, 1]$. The $S_{w,r}$ increase in f strictly, continuously, and (except when $w = r$) singularly, from 0 at $f = 0$ to 1 at $f = 1$. They are definable parametrically, thus: $S(a) = Q_w(a)$, and $f(a) = Q_r(a)$. Each $S_{w,r}$ is related to $S_{r,w}$, $S_{\bar{w},\bar{r}}$, and $S_{\bar{r},\bar{w}}$ by the identities (2.9). For $w = r$, $S_{w,r}(f) = f$. For $w \le r$, $S_{w,r}$ satisfies the full casino inequality (4.2.1). Some special values of S are implicit in Table 5.2.1.

The following remarks are little more than review: $S_{w,t}(S_{t,r}(f)) = S_{w,r}(f)$; for fixed f and r, $S_{w,r}(f)$ is analytic in w in the domain $\{w\colon |w| < 1, |1 - w| < 1\}$.

Study of $S_{1/2,r}(f) = Q_r^{-1}(f)$ as a function of r near $r = f$ shows that $S_{w,r}$ is not always differentiable, let alone analytic, in r.

In common sense, $S_{w,r}(f)$ ought (for $0 < f < 1$) to increase strictly in w from 0 to 1 and decrease strictly in r from 1 to 0. This conclusion is easily verified in steps: first for $Q_w(f)$ with f a binary rational of order 0, then of order n, and then of order ∞; and finally for general w and r. A different and more intuitive argument will be given to illustrate Theorem 8.2.2.

Since, for integral n, $S(r^n) = w^n$, it is suggestive to compare S with the function S' for which this is true for all real n.

$$S'(f) = f^{\log_r w} = f^{(\log w/\log r)} = f^\alpha, \tag{1}$$

with an evident abbreviation. For $w = r$, S' is the same as S, and $\alpha = 1$. If w is less than r, then α exceeds 1 and S' (like S) is a casino function, as was shown in Section 4.2. Therefore, in view of Theorem 7.1, $S'(f)$ majorizes $S(f)$ for $w \le r$ (vice versa for $w \ge r$).

Except when $w = r$, the only solutions of $S'(f) = S(f)$ are $f = r^n$ (for nonnegative integers n) and $f = 0$. These values evidently are solutions. If any other f were a solution, repeated division by r would produce a solution f with $r < f < 1$, which leads to a contradiction, thus. If, for example, $w < r$, then

$$S'(f) = S(f) = w + \bar{w}S\left(\frac{f - r}{\bar{r}}\right) \tag{2}$$

$$\leq wS'(1) + \bar{w}S'\left(\frac{f - r}{\bar{r}}\right)$$

$$< S'\left(r1 + \bar{r}\,\frac{f - r}{\bar{r}}\right) = S'(f).$$

Not only does S' majorize S (for $w < r$), but S' and S are of the same order of magnitude in that $S(f)/S'(f)$ is bounded away from 0 for $f \in (0, 1)$, which is more than is implied by (4.3.14). To see this, together with something else of interest, consider the logarithm of the ratio S/S' as a function of $z = -\log f$, namely,

$$\log S(e^{-z}) + \frac{\log w}{\log r}\, z, \qquad z \geq 0. \tag{3}$$

This function is continuous in z, because it is the logarithm of a positive continuous function, and it is periodic of period $-\log r = |\log r|$. The function (3) is therefore uniformly bounded, which is in effect what was to be proved.

The relation between S' and S has, of course, a dual stemming from (2.9)—specifically, from $S_{w,r}(f) = \bar{S}_{\bar{w},\bar{r}}(\bar{f})$. The whole situation is summarized in the following theorem, which remains true if all the inequalities in it are simultaneously reversed and "min" is replaced by "max".

THEOREM 1. If $w \leq r$,

$$S_{w,r} \leq \min\,(S'_{\alpha}, S''_{\beta}), \tag{4}$$

where

$$S'_{\alpha}(f) = f^{\alpha}, \qquad S''_{\beta}(f) = \overline{\bar{f}^{\beta}} = 1 - (1 - f)^{\beta}, \tag{5}$$

and

$$\alpha = \frac{\log w}{\log r}, \qquad \beta = \frac{\log \bar{w}}{\log \bar{r}}. \tag{6}$$

When $w = r$, $\alpha = \beta = 1$ and $S(f) = S'(f) = S''(f) = f$. When $w < r$, $\alpha > 1 > \beta$; the solutions of $S(f) = S'(f)$ are 0 and r^n; those of $S(f) = S''(f)$ are 1 and $1 - \bar{r}^n$; and the common solutions are 0, r, 1.

When picturing the graphs of S, S'_α, and S''_β it is helpful to know that $S'_\alpha(f) - S''_\beta(f)$ has the same sign as $f - r$ for $w < r$, as is implicit in the proof of the next theorem.

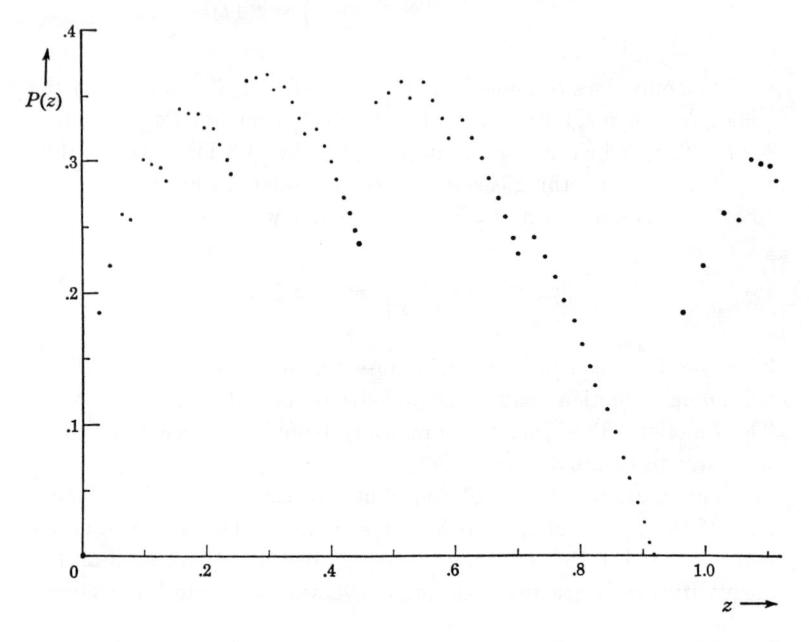

Figure 2 Some points on the graph of $P(z) = -\left[\log S(e^{-z}) + \dfrac{\log w}{\log r} z\right]$ for $(w, r) = (.2, .4)$.

When attention is confined to $w \leq r$ and $w' \leq r'$, the situation of interest, a number of questions suggest themselves as to when $\Gamma_{w,r}(f)$ is at least as favorable, or just as favorable, as $\Gamma_{w',r'}$. These questions are of course expressible in terms of $S_{w,r}(f)$ and $S_{w',r'}(f)$.

In the first place, $S_{w,r}(f) = f$ for all f if and only if $w = r$. If $w < r$, $S_{w,r}(r) = w < r$, and since $S_{w,r}$ is a casino function $S_{w,r}(f) < f$ for all $f \in (0, 1)$, in accordance with the passage beginning Section 4.4. With this remark, let the possibilities $w = r$ and $w' = r'$ be dropped from most of the further discussion.

A necessary and sufficient condition for $S_{w,r}$ to majorize $S_{w',r'}$, but one not always convenient for theoretical problems, is that $S_{w,r}(r') \geq w'$, or, equivalently, that $Q_r^{-1}(r') \geq Q_w^{-1}(w')$. Majorization need not be strict in (0, 1), in that $S_{w,r}(f)$ may be at least $S_{w',r'}(f)$ at all f, and equal to $S_{w',r'}$ at many but not all f.

Example 1. $S_{w,r}$ majorizes S_{w^2,r^2}, and the two functions are equal at $f = r^{2n}$, but $S_{w,r}(r) = w > S_{w^2,r^2}(r)$, in view of Theorem 1.

A sufficient condition for majorization is that $w \geq w'$ and $r \leq r'$. This is perfectly evident from the gambler's viewpoint and is demonstrated by the strict monotony of S in w and r. If either inequality is strict, the majorization is strict in (0, 1).

If $\alpha > \alpha'$, $f^\alpha/f^{\alpha'}$ approaches 0 as f approaches 0. But $S_{w',r'}(f)/f^{\alpha'}$ is bounded away from 0. Therefore, for f sufficiently small, $S_{w',r'}(f) > f^\alpha \geq S_{w,r}(f)$. Dually, if $\beta < \beta'$, $S_{w,r}(f) < S_{w',r'}(f)$ for f sufficiently close to 1. This principle has several applications.

COROLLARY 1. A necessary condition for $S_{w,r}$ to majorize $S_{w',r'}$ is that $\alpha \leq \alpha'$ and $\beta \geq \beta'$.

This condition is not sufficient, since the largest w' compatible with it for given values of w, r, and r', namely, $\min (r'^\alpha, 1 - \bar{r}'^\beta)$, has an isolated corner as a function of r', whereas the largest w' allowed by the known necessary and sufficient condition is singular in r'.

Next, if $\alpha > \alpha'$ and $\beta > \beta'$, the corresponding $S_{w,r}$ and $S_{w',r'}$ cross each other in the sense that the first is smaller than the second for all small f and vice versa for large f.

Example 2. One easy way to find examples of this kind of crossing is to try $w' = \bar{r}$ and $r' = \bar{w}$. Then $w < r$ entails $w' < r'$; and $\alpha' = \beta^{-1}$, $\beta' = \alpha^{-1}$. It remains only to find w and r for which $\alpha\beta > 1$. But with $w = 1/2$ and $\bar{r}$ small, $\alpha\beta$ is rather close to $(\log 2)^2/\bar{r}|\log \bar{r}|$, which is large.

Two primitive casino functions can cross each other infinitely often.

Example 3. Let r be less than r' but $\alpha = \log w/\log r =$

$\log w'/\log r'$. Then $S_{w,r}(f) \leq f^\alpha$, and $S_{w',r'}(f) \leq f^\alpha$, with equality in the first case only at $f = r^n$ and in the second at $f = r'^n$, $n = 1, 2, \cdots$. For typical choices of r and r', these points of contact with f^α are distinct.

Finally, Theorem 1 leads to a demonstration of this anticipated theorem:

THEOREM 2. For all w, r, w', r' in $(0, 1)$. $S_{w,r}$ is identically $S_{w',r'}$ if and only if $w = r$ and $w' = r'$, or $w = w'$ and $r = r'$.

Proof: It is evident that $S_{w,r}$ is $S_{w',r'}$ in the special cases provided for. On the other hand, if $S_{w,r}$ is $S_{w',r'}$, then necessarily $\alpha = \alpha'$ and $\beta = \beta'$, that is,

$$\frac{\log w'}{\log r'} = \alpha, \qquad \frac{\log \bar{w}'}{\log \bar{r}'} = \beta \tag{7}$$

or

$$w' = r'^\alpha, \qquad \bar{w}' = \bar{r}'^\beta. \tag{8}$$

If $w = r$, then $\alpha = \beta = 1$, and $w' = r'$. If $w \neq r$, then $(\alpha - 1)(\beta - 1) < 0$; (8) of course has the solution $w' = w$ and $r' = r$, but if it had another, there would be two different roots r' of

$$r'^\alpha + (1 - r')^\beta - 1 = 0 \tag{9}$$

in the open interval $(0, 1)$. Since (9) is satisfied by $r' = 0$ and $r' = 1$, there would be at least three distinct zeros in $(0, 1)$ of the derivative of the left side of (9) with respect to r', that is, at least three distinct roots of

$$\alpha r'^{(\alpha-1)} = \beta(1 - r')^{\beta-1}, \tag{10}$$

or, since β is neither 0 or 1, of

$$(1 - r') = \left[\frac{\alpha r'^{(\alpha-1)}}{\beta}\right]^{1/(\beta-1)}. \tag{11}$$

This is impossible, for the left side of (11) is linear and the right side is strictly convex. ◆

THEOREM 3. For $0 < w < r < 1$, $S_{w,r}(\bar{w}) \leq \bar{r}$ with strict inequality unless $w + r = 1$. In particular, for $0 < w < 1/2$, $Q_w(\bar{w}) < 1/2$.

Proof: $S_{w,r}$ and $S_{\bar{r},\bar{w}}$ are dual functions according to the remark following (2.9). If $w + r = 1$, and, according to Theorem 2, only then, $S_{w,r}$ is self-dual. Otherwise, neither $S_{w,r}$ nor $S_{\bar{r},\bar{w}}$ majorizes the other, in accordance with the observation near the end of Section 4.6. Apply Corollary 7.2 to conclude that $S_{w,r}(\bar{w}) < \bar{r}$. ◆

As is easy to see, the *w-coin distribution* γ_w whose distribution function is Q_w has $\bar{w}$ as its mean. Therefore the final sentence of Theorem 2 can be interpreted thus:

COROLLARY 2. For $0 < w < 1/2$, the mean of the w-coin distribution γ_w is less than its median.

Of course the identities (2.9) and Theorem 3 imply that $S_{w,r}(\bar{w}) > \bar{r}$ if $0 < r < w < 1$. In particular, $Q_w(\bar{w}) > 1/2$ for $1/2 < w < 1$.

COROLLARY 3. $Q_w(\bar{w}) < Q_{\bar{w}}(w)$ for $0 < w < 1/2$.

Corollary 3 can be interpreted as a partial answer to whether, in bold play at red-and-black, it is preferable to have a large probability of winning each individual bet with a small initial fortune or a small probability of winning each bet with a large initial fortune.

It is now easy to refine Theorem 7.1.

THEOREM 4. If U is a casino function (or even a bounded, nonzero solution of (4.3.1) and (4.3.2)) and $w < U(r)$, then $S_{w,r}(f) < U(f)$ for all f in $(0, 1)$. If $w \leq U(r)$, then $S_{w,r} \leq U$.

Proof: Choose w' satisfying $w < w' < U(r)$. Then $S_{w,r}(f) < S_{w',r'}(f) \leq U(f)$ for f in $(0, 1)$. The first inequality is the strict monotony of S in w, and the second is asserted in Theorem 7.1. ◆

9. UNIFORM ROULETTE

A roulette table, like a primitive casino, is governed by two parameters w and r. In Nevada, w and r are 1/38 and 1/36 (Scarne

1961, p. 362). At Monte Carlo, w and r are 1/37 and 1/36, and the situation is more complicated (Scarne 1961, p. 356). Because of this disparity between Nevada and Monte Carlo rules and, more important, because the problem of optimal play at a roulette table remains open, there is no point in giving precise definitions here. If the bets open to the gambler are restricted somewhat, however, optimal strategies are easy to find and demonstrate, using knowledge of primitive casinos. A *uniform-roulette table* $\hat{\Gamma}_{w,r}$ is the union of the primitive casinos $\Gamma_{w,r}, \Gamma_{2w,2r}, \cdots, \Gamma_{nw,nr}$, where n is the largest integer for which $nr < 1$. When a gambler plays red-and-black in Nevada, he is, in effect, selecting a gamble from $\Gamma_{18w,18r}$, where w and r are 1/38 and 1/36. More generally, if on a spin of the wheel he stakes s on each of j numbers, he is in effect selecting a gamble from $\Gamma_{jw,jr}$. A gambler at a uniform-roulette table may vary s and j on different spins. If on some spin of the wheel he stakes different positive amounts $s_1, \cdots, s_j$ on j numbers, then his play is no longer uniform. A uniform-roulette table is plainly a casino in the technical sense of Chapter 4.

THEOREM 1. If $0 < w < r < 1$, and $0 < f < 1$, then $S_{w,r}(f) > S_{jw,jr}(f)$ for all integers $j > 1$ for which $jr < 1$.

Proof: According to Theorem 5.1, $S_{w,r}(jr) > jS_{w,r}(r) = jw = S_{jw,jr}(jr)$. Now apply Theorem 8.4. ♦

THEOREM 2. The utility of the uniform-roulette casino $\hat{\Gamma}_{w,r}$ is the primitive casino function $S_{w,r}$. The only gambles in $\hat{\Gamma}_{w,r}(f)$ that conserve $S_{w,r}$ at f are in $\Gamma_{w,r}(f)$.

Proof: Apply Theorem 1 and Theorem 7.2. ♦

As Theorem 2 implies, the bold strategy based on the longest odds is optimal in a uniform-roulette casino. It is natural to conjecture that this strategy is also optimal for not necessarily uniform roulette, but this remains unsettled.

THEOREM 3. For all real j for which $0 < w < r < jr < 1$, $S_{jw,jr}(r)$ is strictly less than w.

Proof: $S_{jw,jr}(r) = (jw)S_{jw,jr}(j^{-1}) < (jw)(j^{-1}) = w$. ♦

One might conjecture that it is not important for j in Theorem 1 to be an integer. Indeed, Theorem 3 implies that $S_{jw,jr}$ cannot majorize $S_{w,r}$ if $j > 1$. Nevertheless, $S_{w,r}$ need not majorize $S_{jw,jr}$ either.

Example 1. Let $r = 1/2$, $j = 4/3$, and w be arbitrary in $(0, 1)$. According to Table 5.2.1, $S_{w,r}(2/3) = S_{w,1/2}(2/3) = Q_w(2/3) = w(1 - w\bar{w})^{-1}$. Consequently, for $w \neq 1/2$, $S_{w,r}(2/3) < 4w/3 = S_{4w/3,2/3}(2/3)$. It seems noteworthy that, for all $w < 1/2$, Q_w and $S_{4w/3,2/3}$ assume equal values at some point in the interval $(1/2, 2/3)$.

For $j > 1$, the α of $\Gamma_{jw,jr}$ is greater, and the β smaller, than the α and β of $\Gamma_{w,r}$. Therefore, Example 1 illustrates concretely that the condition of Corollary 8.1 is not sufficient.

According to Example 1, min $\{S_{tw,tr}(f)\colon 0 < t < 1\}$, say $\hat{S}_{w,r}(f)$, can be strictly less than $S_{w,r}(f)$, at least for $r = f = 2/3$. As Theorem 4.6.1 implies, $\hat{S}_{w,r}$ is a casino function. Knowledge of other properties of this function is desirable—in particular, knowledge of interesting lower bounds for $\hat{S}_{w,r}(r)$.

7

SOME GENERAL PRINCIPLES

1. INTRODUCTION. This chapter is about some general principles that apply to gambling houses on any set F of fortunes, and the next is about some general principles that apply only when F is a set of real numbers.

As a preliminary, certain facts and technical terms so natural that they might almost go unmentioned are catalogued. Some of them have been encountered in earlier sections.

If Γ_1 and Γ_2 have the same space of fortunes F and $\Gamma_1(f) \supset \Gamma_2(f)$ for all f in F (or $\Gamma_1 \supset \Gamma_2$, for short), then Γ_1 is a *superhouse* of Γ_2, and Γ_2 is a *subhouse* of Γ_1. If $\Gamma_1 \supset \Gamma_2$, the same relation subsists between the leavable closures Γ_1^L and Γ_2^L. If $\Gamma_1 \supset \Gamma \supset \Gamma_2$ or $\Gamma_2 \supset \Gamma \supset \Gamma_1$, then Γ is *between* Γ_1 and Γ_2.

The union $\bigcup_\alpha \Gamma_\alpha$ of a nonvacuous class of gambling houses on a fixed F is defined by

$$(1) \qquad (\bigcup_\alpha \Gamma_\alpha)(f) = \bigcup_\alpha \Gamma_\alpha(f)$$

for all f in F. It is a superhouse of each Γ_α, and $(\bigcup_\alpha \Gamma_\alpha)^L = \bigcup_\alpha \Gamma_\alpha^L$. Typically, many strategies available in a union are not available in any one term of the union.

"Intersection" and "$\bigcap$" can almost be substituted for "union" and "$\bigcup$" in the preceding paragraph, but some differences must be mentioned. Since we have chosen to require, for a gambling house Γ, that $\Gamma(f)$ be nonvacuous, the intersection of a class of gambling houses can fail to be a gambling house. It always is a gambling house if the elements of the class are leavable, and the intersection of a class is leavable if and only if the elements of the class are. A strategy is available in $\bigcap_\alpha \Gamma_\alpha$ at f if and only if it is available in each Γ_α at f.

2. COMPOSITION

Though the gambling house Γ permits the gambler with fortune f to use immediately only the gambles in $\Gamma(f)$, the gambler can, in effect, generally build up or compose other gambles for himself in the course of time. Since time here is at no premium, these composite gambles are, to all intents and purposes, as available to the gambler as are the elements of $\Gamma(f)$ themselves. If, for example, $\gamma \in \Gamma(f)$ and $\gamma_g \in \Gamma(g)$, then the new gamble γ' for which each bounded real-valued v has the expectation

$$\gamma' v = \int \gamma_g v \, d\gamma(g), \tag{1}$$

is effectively available at f. (This γ' is the distribution of the terminal fortune of an available two-move policy with initial fortune f. The expectation of v under γ' is the expectation under γ of the conditional expectation of v under the conditional distribution γ_g.)

More generally, if $\pi = (\sigma, t)$ is a policy available to the gambler at f, he can compose for himself $\gamma(\pi)$, the distribution of f_t under σ. Recapitulating (2.9.4),

$$\gamma(\pi)v = \sigma v(f_t) \tag{2}$$

for all bounded functions v.

Possibly the operation (2), not to mention its special case (1), does not actually augment $\Gamma(f)$ for any f. That is, perhaps $\gamma(\pi) \in \Gamma(f)$. In this case, Γ is *closed under composition*. Even the γ' constructed according to (1) provide an adequate test of closure under composition, as can be shown easily with or without the help of (2.9.7):

THEOREM 1. If, for each f, every γ', in the sense of (1), is included in $\Gamma(f)$, then Γ is closed under composition.

A simple example of a gambling house Γ_v closed under composition is constructed from any bounded, real-valued v with F as its domain by letting

$$(3) \qquad \Gamma_v(f) = \{\gamma: \gamma v \leq v(f)\} = \{\gamma: v \text{ is excessive for } \gamma \text{ at } f\};$$

this is the *full house under* v. Reversal of the inequality in (3) and replacement of the inequality by equality also lead to houses closed under composition—the full house under $-v$ and the *v-conserving house.* The important special case of a house conserved by an exponential function was introduced and exploited in (de Finetti 1939) and is studied in Section 8.7.

An intersection of any number of houses closed under composition is itself closed under composition. In particular, the intersection of any class of full houses is closed under composition and is leavable. A conserving house, for instance, can be viewed as the intersection of two full houses. Full houses will be discussed frequently again.

Any house Γ determines its *composition closure* Γ^c, the intersection of all superhouses of Γ that are closed under composition. The composition closure of Γ can be defined much more constructively. Let $\Gamma''(f) = \{\gamma(\pi): \pi \text{ is available in } \Gamma \text{ at } f\}$, where $\gamma(\sigma, t)$ is the distribution under σ of f_t as in (2).

THEOREM 2. $\Gamma^c = \Gamma''$.

Proof: Plainly, $\Gamma^c \supset \Gamma''$. The reverse inclusion holds if Γ'' is closed under composition. In view of Theorem 1, it is therefore necessary only to show that $\gamma''' \in \Gamma''(f)$, where $\gamma''' v = \int \gamma_g'' v \, d\gamma''(g)$, whenever $\gamma'' \in \Gamma(f)$ and $\gamma_g'' \in \Gamma''(g)$ for all g. This is done with the help of (2.9.7). ◆

If Γ is closed under composition, then $\Gamma = \Gamma^c$, and whatever can be achieved by a policy available at f can also be achieved by a single γ available at f, which sometimes (as in Section 9.2) simplifies the evaluation of U.

THEOREM 3. If Γ is leavable and closed under composition, then $U(f) = V(f) = \sup \gamma u$ over γ in $\Gamma(f)$.

3. PERMITTED SETS OF GAMBLES

A function v from fortunes to real numbers determines a set v° of gambles,

$$v^\circ = \{\gamma\colon \gamma v \leq 0\}, \tag{1}$$

the gambles *permitted* by v. Such permitted sets of gambles are introduced here because of the importance of some *permitted houses*, that is, houses in which each $\Gamma(f)$ is the set of gambles permitted by some v_f.

A full house Γ_v in the sense of (2.3) is, of course, the permitted house with $v_f = v - v(f)$, and these are the most important examples of permitted houses. Later sections will have much to say about full houses and a little about other permitted houses. The present section concerns mainly the relation between two functions v and w when the permitted set v° is a subset of w°, or w is *(at least) as permissive as* v.

Since unbounded functions v are often of interest, a clear and appropriate convention is needed for the meaning of the inequality $\gamma v \leq 0$. For this and later purposes, γv is extended to certain unbounded v by the following familiar sequence of definitions:

If v is nonnegative, $\gamma v = \lim (\gamma \min (v, c))$ as the number c approaches ∞; with ∞ admitted as a possible value, γv always exists for nonnegative v. If v is nonpositive, $\gamma v = -\gamma(-v)$. For an arbitrary v, $\gamma v = \gamma v^+ + \gamma v^-$, provided at least one of the two terms is finite, where, as usual, $v^+ = \max (v, 0)$ and $v^- = \min (v, 0)$.

In particular, $\gamma v \leq 0$ if and only if $\gamma v^+ < \infty$ and $\gamma v^+ \leq |\gamma v^-|$, where $|\gamma v^-|$ may be ∞.

THEOREM 1.

(a) $\gamma v \leq 0$ for all γ (that is, every gamble belongs to v°) if and only if $v \leq 0$.

(b) $\gamma v \leq 0$ for no γ (that is, v° is vacuous) if and only if v is uniformly positive (that is, there is an ϵ for which $v \geq \epsilon > 0$).

(c) $\delta(g) \in v^\circ$ if and only if $v(g) \leq 0$.

(d) If f^+ and f^- are fortunes for which $v(f^+) > 0$ and $v(f^-) < 0$, then the two-point gamble γ for which

$$\gamma\{f^+\} = \frac{-v(f^-)}{v(f^+) - v(f^-)}, \qquad \gamma\{f^-\} = \frac{v(f^+)}{v(f^+) - v(f^-)} \tag{2}$$

belongs to v°, and $\gamma v = 0$.

(*e*) If v is unbounded from below, there is, for every f, a two-point gamble γ in v° with $\gamma\{f\}$ arbitrarily close to 1.

(*f*) If v is bounded from below, $z = \inf v \leq 0$, and $v(g) > 0$, then

$$\sup_{\gamma \in v^\circ} \gamma\{g\} = \frac{-z}{v(g) - z}, \tag{3}$$

and, for all $\gamma \in v^\circ$ and $\epsilon > z$,

$$\gamma\{f: v(f) \geq \epsilon\} \leq \frac{-z}{\epsilon - z}. \tag{4}$$

Of course, each part of Theorem 1 has practical implications for a gambler constrained to choose a gamble from v°. For example, part (*e*) shows that, if v is unbounded from below and the gambler's utility u is bounded, then the gambler can practically have his heart's desire with a single two-point gamble in v°. Part (*f*) bounds what he can achieve in v° when v is bounded from below.

If $v \leq aw$ for some nonnegative constant a, then $\gamma w \leq 0$ plainly implies that $\gamma v \leq 0$; in short, $v^\circ \supset w^\circ$, or v is as permissive as w. Similarly, if $v = aw$ for some positive a, then $v^\circ = w^\circ$; v and w permit the same gambles. Several paragraphs lead to the next two theorems asserting almost the converses of these facts.

Evidently, from part (*a*) of Theorem 1, if $w \leq 0$, then $v^\circ \supset w^\circ$ if and only if $v \leq 0$. The condition that $v \leq 0$ can, though artificially, be written $v \leq 0w$.

Suppose, for completeness, that w is nowhere negative. Then $\gamma w \leq 0$ if and only if $\gamma\{f: w(f) > \delta\} = 0$ for each positive δ. If for each positive ϵ there is a positive δ such that $v(f) > \epsilon$ implies $w(f) > \delta$, then $\gamma w \leq 0$ implies $\gamma v \leq 0$, or $v^\circ \supset w^\circ$. Conversely, if, for some f, $v(f) > 0$ and $w(f) = 0$, $\delta(f)$ is in w° but not in v°; similarly, if there is a sequence of distinct fortunes f_i with $w(f_i) \leq i^{-1}$ but $v(f_i) > \epsilon > 0$, a diffuse gamble on the sequence $\{f_i\}$ belongs to w° but not to v°. In summary, if w is nowhere negative, then v is as permissive as w if and only if, for every positive ϵ, there is a positive δ such that $v(f) > \epsilon$ implies $w(f) > \delta$. (When discussion is restricted to countably additive γ on some sigma-field with respect to which v and w are measurable, it is necessary and sufficient that $v(f) > 0$ imply $w(f) > 0$.)

Example 1. On $[0, 1]$, let $u(f) = f(1 - f)$, $v(f) = f^2$, and $w(f) = 0$ or 1 according as f is 0 or positive. Then $u^\circ \supset v^\circ \supset w^\circ$, and neither inclusion can be reversed (though, in a countably additive setting, the second one could be). In this example there is no nonnegative a for which $u \leq av$.

Suppose, finally, that the two sets of fortunes, F^+ where $w(f)$ is positive and F^- where it is negative, are both nonvacuous. For any fortunes f^+ and f^- in F^+ and F^-, consider that two-point gamble γ carried by those two fortunes for which $\gamma w = 0$. A necessary condition for $v^\circ \supset w^\circ$ is that $\gamma v \leq 0$. Equivalently, $-w(f^-)v(f^+) + w(f^+)v(f^-) \leq 0$, or

$$\frac{v(f^+)}{w(f^+)} \leq \frac{-v(f^-)}{-w(f^-)}. \tag{5}$$

If these necessary conditions obtain for all $f^+ \in F^+$ and $f^- \in F^-$, then the range of neither side of (5) is vacuous, that of the right side is nonnegative, and there is some nonnegative number a between the supremum of the range of the left side and the infimum of the range of the right side. For such an a, $v(f) \leq aw(f)$ if $w(f) \neq 0$. If $w(f) = 0$, $\delta(f)w = 0$, and $\delta(f)v \leq 0$ implies $v(f) \leq 0$. So $v \leq aw$. Consequently, the condition that $v \leq aw$, which is always sufficient for $v^\circ \supset w^\circ$, is, in this case, also necessary.

Condensing the three possibilities:

THEOREM 2. $\gamma w \leq 0$ implies $\gamma v \leq 0$, or v is as permissive as w, if and only if one of the following two conditions obtains:

(*a*) w is nowhere negative, and for every positive ϵ there is a positive δ such that $v(f) > \epsilon$ implies $w(f) > \delta$.

(*b*) w is somewhere negative, and there is a nonnegative a such that $v(f) \leq aw(f)$ for all f.

(The hypothesis of Theorem 2 says that no linear functional separates v from the cone generated by w and all nonpositive functions, so that v is in the "natural closure" of the cone, and part (*b*) says that then v must be in the cone itself. However, a rigorous formulation and proof along this line would not fall under the most elementary general theories; the need for part (*a*) seems symptomatic of a need for special handling.)

Except for two extreme and not very important cases, $v^\circ = w^\circ$ if and only if v and w are positive multiples of each other:

THEOREM 3. $\gamma v \leq 0$ is equivalent to $\gamma w \leq 0$, or $v^\circ = w^\circ$, if and only if one of the following three conditions obtains:

(*a*) v and w are nowhere positive.

(*b*) Condition (*a*) of Theorem 2 applies in both directions.

(*c*) For some positive number a, $v = aw$.

Proof: If v or w is nowhere positive, part (*a*) of Theorem 1 establishes condition (*a*) of the present theorem.

If v and w are nowhere negative, Theorem 2 implies condition (*b*) of the present theorem.

If w, say, is somewhere positive and somewhere negative and $v^\circ = w^\circ$, then $v \leq aw$ for some nonnegative a, according to Theorem 2. The possibility that v is nowhere positive is excluded, and so a is positive. The possibility that v is nowhere negative is now also excluded; so, according to Theorem 2, $w \leq bv$ for some nonnegative b. Since $v \leq aw \leq abv$ and since $v(f)$ can be positive and can be negative, $ab = 1$, and $v = aw$. ◆

It is of some interest for gambling problems to determine, for a bounded function u, the supremum of γu over the γ permitted by a function w. This is plainly the least number b (or $-\infty$ if and only if w is uniformly positive) for which $\gamma u \leq b$ for all γ in w°, that is, the least number b for which $(u - b)$ is as permissive as w. According to Theorem 2, there are two cases to consider.

First, suppose $w \geq 0$. For which numbers b' does $(u - b')$ satisfy condition (*a*) of Theorem 2? For those such that for each positive ϵ there is a positive δ such that wherever $u(f) > b' + \epsilon$, $w(f) > \delta$; that is, numbers as large as

$$\inf_{\delta > 0} \sup_{w(f) < \delta} u(f). \tag{6}$$

The number (6) is therefore b in this case. If competition is restricted, in a countably additive spirit, to gambles for which $\gamma w \leq 0$ implies that $\gamma\{f\colon w(f) > 0\} = 0$ (as when γ is discrete), then (6) is replaced by

$$\sup_{w(f) = 0} u(f). \tag{7}$$

And this value can obviously be approached by means of one-point gambles.

Second, if w is somewhere negative, then b is the least number for which there is a nonnegative number a such that $u \leq aw + b$. The discussion on which Theorem 2 is based shows that the value of b in this case is not affected by restricting competition to one-point and two-point gambles.

The description of b in the second case actually applies in the first case as well. The b thus described is obviously large enough. To see that it is not too large, suppose that $w \geq 0$ and b' exceeds the number (6). There is a δ such that, wherever $u(f) > b'$, $w(f) > \delta$. Suppose $a \geq \sup (u/\delta)$. Then $u \leq aw + b'$.

THEOREM 4. For a bounded u, sup γu over γ in w° is the least b such that, for some nonnegative number a, $u \leq aw + b$. If w is nonnegative, (6) is an alternative evaluation. In this special case, (6) must be replaced by (7) if the competition is restricted to gambles for which $\gamma\{f: w(f) > 0\} = 0$; and this latter bound remains valid even if the competition is restricted still further to one-point gambles. Except in the special case, neither the sup γu nor b is decreased if competition is restricted to one-point and two-point gambles.

The following modifications of Theorems 2, 3, and 4 are left to the interested reader.

THEOREM 5. $\gamma w = 0$ implies $\gamma v \leq 0$ if and only if one of the following two conditions obtains:

(*a*) w is nowhere negative or nowhere positive, and for every positive ϵ there is a positive δ such that $v(f) > \epsilon$ implies $|w(f)| > \delta$.

(*b*) w is somewhere positive and somewhere negative, and there is a real number a for which $v \leq aw$.

If v is nonnegative (positive) at some f where w is positive, then a in condition (*b*) must be nonnegative (positive); if v is nonpositive wherever w is negative, then a can be nonnegative. Therefore, if $v(f^+) \geq v(f^-)$ wherever $w(f^+) > 0 > w(f^-)$, then a can be nonnegative; so if v is of the form $v'(w)$ with v' nondecreasing on the range of w, then a can be nonnegative.

COROLLARY 1. $\gamma w = 0$ implies $\gamma v = 0$ if and only if one of the following two conditions obtains:

(*a*) w is nowhere negative or nowhere positive, and for every positive ϵ there is a positive δ such that $|v(f)| > \epsilon$ implies $|w(f)| > \delta$.

(*b*) w is somewhere positive and somewhere negative, and there is a real a for which $v = aw$.

COROLLARY 2. $\gamma w = 0$ is equivalent to $\gamma v = 0$ if and only if condition (*a*) of Corollary 1 applies in both directions or condition (*b*) of Corollary 1 applies with a unequal to 0.

Every part of Theorem 4 can be easily adapted to the set of gambles for which $\gamma w = 0$, but only the main counterparts are explicitly expressed in the next theorem.

THEOREM 6. For a bounded u, sup γu over $\{\gamma\colon \gamma w = 0\}$ is the least b such that, for some real number a, $u \leq aw + b$. If u is a nondecreasing function of w, then it is enough to admit nonnegative values of a. If w is nonnegative or nonpositive, (6) applied to $|w|$ in place of w is an alternative evaluation of sup γu (whether u is a nondecreasing function of w or not).

4. FULL HOUSES

According to (2.3), Γ_v, the full house under v, is the house for which $\Gamma_v(f) = (v - v(f))^\circ$. The assumption in Section 2 that v is bounded was made for momentary simplicity only and is now dropped. Clearly, Γ_v is the largest house for which v is excessive. If both v and $-v$ are excessive for a house Γ so that $\gamma v = v(f)$ for all f and all $\gamma \in \Gamma(f)$, then Γ *conserves* v. The largest house that conserves v, the intersection of Γ_v and Γ_{-v}, is the v-conserving house defined in Section 2.

If v is excessive for Γ, then all $av + b$, with b real and a positive, also are. Therefore, v and $av + b$ define the same full and conserving houses. The converse is also true.

THEOREM 1. The full house under v is the full house under w if and only if

$$w = av + b \tag{1}$$

for some real b and some positive real a. The v-conserving house is the w-conserving house if and only if (1) holds for some real a and b with $a \neq 0$.

Proof: As already mentioned, the "if" part of the first assertion is obvious. To prove the "only if" part, consider first the situation in which v or w takes on at least three values. There is no further loss of generality in assuming that, for some f, $v - v(f)$ is both somewhere positive and somewhere negative. As follows immediately from Theorem 3.3, if $\Gamma_v = \Gamma_w$, then

$$w - w(f) = a(v - v(f)) \tag{2}$$

for some positive a. This is equivalent to (1) and disposes of the situation in which v or w takes on at least three values. The remaining possibilities for the first assertion are trivial. Proof of the assertion about conserving houses is left to the reader. ◆

The assertion of Theorem 1 about conserving houses is close to (Hardy, Littlewood, and Polya 1934, Theorem 83, p. 66).

Two simple principles:

THEOREM 2. If $u \leq av + b$, then $U \leq av + b$ for any subhouse Γ of the v-conserving house and, provided $a > 0$, for any subhouse of the full house under v as well.

Proof: Immediate from the basic Theorem 2.12.1. ◆

THEOREM 3. If u coincides with $av + b$ at two fortunes, say f and h, and if Γ is a superhouse of the v-conserving house, then $U(g) \geq av(g) + b$ at every g for which $av(g) + b$ is between $u(f)$ and $u(h)$.

Proof: The assertion is trivial unless the numbers $u(f)$, $u(h)$, and $av(g) + b$ are all different, in which case there is no loss of generality in assuming $u(f) < av(g) + b < u(h)$. Then, according to part (d) of Theorem 3.1, there is a v-conserving two-point gamble γ in $\Gamma(g)$ carried on f and h such that $\gamma u = a\gamma v + b = av(g) + b$, which proves the assertion. Explicitly,

$$\gamma\{f\} = \frac{v(h) - v(g)}{v(h) - v(f)} = \frac{u(h) - [av(g) + b]}{u(h) - u(f)}; \tag{3}$$

$$\gamma\{h\} = \frac{v(g) - v(f)}{v(h) - v(f)} = \frac{[av(g) + b] - u(f)}{u(h) - u(f)}. \tag{4}$$

◆

Example 1. Let v be any function on $[0, 1]$ that attains its minimum at 0 and its maximum at 1, and let $u(1) = 1$ and $u(f) = 0$ for f in $[0, 1)$. According to Theorem 1, no generality is lost in assuming that $v(0) = 0$ and $v(1) = 1$. According to the first principle, U is at most v for any subhouse of the full house Γ; according to the second, this bound is achievable, even in the v-conserving house, by one-move, two-point policies.

Example 2. Let v be defined on $(-\infty, 0]$, with $0 \leq v \leq 1$, $v(0) = 1$, and with $\inf v = 0$. This contrasts with Example 1 only in that a partially diffuse gamble may be needed where a two-point gamble could be used before. If partially diffuse gambles are excluded, there remain arbitrarily good one-move, two-point policies; there are still optimal strategies but no optimal policies.

Example 3. On the interval $[0, 10]$, let v be convex and increasing, and let $u(f)$ be the largest integer that is at most f. For any Γ between the v-conserving house and the full house under v and for f in $[n, n + 1]$ $(n = 0, \cdots, 9)$,

$$U(f) = \frac{n[v(n+1) - v(f)] + (n+1)[v(f) - v(n)]}{v(n+1) - v(n)}, \tag{5}$$

as the two principles show.

This example, unusual in this book in that it solves a problem with a many-valued utility u, admits a generalization that, in a sense, solves all gambling problems for conserving and full houses. The rest of this section is about this generalization (not used in later sections). Theorems 2.3, 3.4, and 3.6 are the tools for the generalization.

THEOREM 4. If Γ is the full house under v, then

$$U(f) = \inf (av(f) + b)$$

over those numbers a and b for which $a \geq 0$ and $av + b \geq u$. At any f at which v attains its minimum,

$$U(f) = \inf_{\delta > 0} \sup_{v(g) < v(f) + \delta} u(g). \tag{6}$$

THEOREM 5. If Γ is the v-conserving house, then

$$U(f) = \inf (av(f) + b)$$

over those numbers a and b for which $av + b \geq u$. If u is a nondecreasing function of v, then it is enough to admit nonnegative values of a. At any f at which v attains its maximum or minimum,

$$U(f) = \inf_{\delta > 0} \sup_{|v(g) - v(f)| < \delta} u(g). \tag{7}$$

It is easy to see how (6) and (7) have to be modified if Γ is replaced by its subhouse of one-point and two-point gambles.

5. PERMITTED HOUSES

Some interesting gambling houses are constructed from a family of functions v_f from the fortunes to real numbers, one function for each f, by letting $\Gamma(f) = \{\gamma: \gamma v_f \leq 0\}$. In the terminology of Section 3, Γ is a permitted house, and $\Gamma(f) = v_f^\circ$. Full houses are interesting examples of permitted houses, as are income-tax casinos (defined in Section 9.2), but later sections rely so little on the present section that some may prefer to skip or defer it.

Practically any family v_f permits some gambling house Γ. What is needed is simply that no v_f be uniformly positive so that every v_f° is nonempty. However, $v_f(f)$ must be nonpositive if Γ is to be leavable.

Immediately from Theorem 3.3:

THEOREM 1. If Γ is permitted by v_f and v_f assumes both positive and negative values for each f, then Γ is also permitted by v_f' if and only if $v_f' = a(f)v_f$ for some positive $a(f)$.

The solution of gambling problems often hinges on finding appropriate excessive functions.

THEOREM 2. Suppose that Γ is permitted by v_f, $a(f) \geq 0$, and

$b(f)$ is real. If, for all f and g, $b(g) \leq a(f)v_f(g) + b(f)$, then b is excessive for Γ.

Theorem 2 will be illustrated in Section 9.2. We are largely in the dark as to when nontrivial functions a and b exist.

A permitted house can be closed under composition, as has already been illustrated by the case of full houses.

THEOREM 3. If $v_f - v_f(g)$ is at least as permissive as v_g for all f and g, then the house permitted by the family v_f is closed under composition.

Proof: Suppose $\gamma v_f \leq 0$ and $\gamma_g v_g \leq 0$ for all g. Under these assumptions and the hypothesis of the theorem,

$$\gamma' v_f = \int \gamma_g v_f \, d\gamma(g) \leq \int v_f(g) \, d\gamma(g) \leq 0. \tag{1}$$

Now apply Theorem 2.1. ♦

COROLLARY 1. If $v_f(f) \leq 0$ and $v_f - v_f(g)$ is at least as permissive as v_g for all f and g,

$$U(f) = V(f) = \sup_{\gamma \in \Gamma(f)} \gamma u. \tag{2}$$

Under a mild hypothesis, Theorem 3 has a converse.

THEOREM 4. If each v_f assumes both positive and negative values and if $v_f(f) \leq 0$ for all f, then the house Γ permitted by v_f is closed under composition if and only if $v_f - v_f(g)$ is at least as permissive as v_g for all f and g.

Proof: The "if" clause is more than covered by Theorem 3.

As for the "only if" clause, let γ be a two-point gamble carried by, say, h and g, such that $\gamma v_f = 0$ and $\gamma\{g\} > 0$; such a gamble exists for an arbitrary f and g under the preliminary hypotheses of the theorem. Let $\gamma_e = \delta(e)$ if $e \neq g$, and let γ_g be any gamble for which $\gamma_g v_g \leq 0$. Thus $\gamma \in \Gamma(f)$ and $\gamma_e \in \Gamma(e)$ for all e. If Γ is closed under composition, then $\gamma' = \int \gamma_e \, d\gamma(e)$ is an element of $\Gamma(f)$, and

$$
\begin{aligned}
(3) \qquad 0 > \gamma' v_f &= \int \gamma_e v_f \, d\gamma(e) \\
&= v_f(h)\gamma\{h\} + (\gamma_g v_f)\gamma\{g\} \\
&= -v_f(g)\gamma\{g\} + (\gamma_g v_f)\gamma\{g\} \\
&= (-v_f(g) + \gamma_g v_f)\gamma\{g\}.
\end{aligned}
$$

Since $\gamma\{g\} > 0$, $\gamma_g(v_f - v_f(g)) = -v_f(g) + \gamma_g v_f \leq 0$. ♦

Theorem 3.4 implies the final theorem and evident refinements of it.

THEOREM 5. If the house permitted by v_f is leavable and closed under composition, $U(f)$ is the infimum of those numbers b for which $u \leq av_f + b$ for some nonnegative a.

8

HOUSES ON THE REAL LINE

1. INTRODUCTION. This chapter studies certain properties of gambling houses for which the space of fortunes F is a set of real numbers. Those who prefer to proceed from the specific to the general may prefer to read the chapter backwards or to look first through its last section, which is largely independent of the earlier ones.

2. DOMINATION

In this section, F is assumed to be a subset of $(-\infty, \infty)$. If γ is a gamble on F and S is any subset of $(-\infty, \infty)$, γS will mean $\gamma(F \cap S)$. If $\gamma v \geq \gamma' v$ for every bounded, nondecreasing, real-valued function v on F, then γ *dominates* γ'.

THEOREM 1. γ dominates γ' if and only if, for all f in F, $\gamma[f, \infty) \geq \gamma'[f, \infty)$ and $\gamma(f, \infty) \geq \gamma'(f, \infty)$.

Pairs of gambles that dominate each other are closely equivalent for many purposes.

COROLLARY 1. γ and γ' dominate each other if and only if they agree on all intervals.

COROLLARY 2. If γ dominates γ' and $\gamma v < \infty$ for a nondecreasing but not necessarily bounded v, then $\gamma' v$ exists and $\gamma' v \leq \gamma v$.

If Δ is a set of gambles and γ is a gamble that some element of Δ dominates, then Δ *dominates* γ. The set of gambles dominated by Δ clearly includes Δ and can be Δ itself, as in the following corollary (where v° is as in (7.3.1)).

COROLLARY 3. If v is nondecreasing, the set of all gambles dominated by v° is v° itself.

If a set of gambles Δ dominates every element of a set Δ^*, then Δ *dominates* Δ^*. If $\Gamma(f)$ dominates $\Gamma^*(f)$ for all f, then Γ *dominates* Γ^*.

THEOREM 2. If Γ dominates Γ^*, then every nondecreasing function v excessive for Γ is excessive for Γ^*. If U is nondecreasing, then U is excessive for Γ^* and majorizes U^*.

Proof: Apply Theorems 2.12.1 and 2.14.1. ◆

(Theorem 2 suggests an argument for the strict monotony of the primitive-casino functions $S_{w,r}(f)$—increasing in w and decreasing in r for $0 < f < 1$—more natural than the combinatorial argument sketched in the paragraph preceding (6.8.1). Theorem 2 yields monotony in the right direction immediately; that this monotony is strict in w follows from (5.2.1); finally, the conclusion for r is immediate from that for w.)

If $\Gamma(f)$ dominates $\Gamma(f')$ whenever $f > f'$, then Γ *increases (not necessarily strictly) in dominance.* As would be expected, if Γ increases in dominance and u is nondecreasing, a rich gambler can do as well as a poor one—at least if Γ is leavable:

THEOREM 3. If u is nondecreasing and Γ increases in dominance, then U_n, U_ω, and U are nondecreasing.

Proof: In the notation introduced parenthetically in Section 2.15, $U'_1 = u$ is nondecreasing, and if U'_β is nondecreasing for $\beta < \alpha$, then U'_α is nondecreasing. By transfinite induction, U'_α is nondecreasing for all α, whence U is nondecreasing. ◆

COROLLARY 4. If Γ dominates Γ^*, Γ increases in dominance, and u is nondecreasing, then U majorizes U^*.

(Under the hypotheses of Corollary 4, U_n and U_ω majorize U_n^* and U_ω^*. But the composition closure of Γ need not dominate that of Γ^*, as easy examples show. It would do so under the following slightly weaker but less simple definition: Δ "dominates" γ if, for each positive ϵ, there is a γ' in Δ such that $\gamma' v \geq \gamma v - \epsilon$ for all nondecreasing v bounded between 0 and 1. Substituting this alternative definition, Corollary 3 and Theorem 2 remain true for v bounded from below, and—strengthening the original assertions—Theorem 3 and Corollaries 4 and 5 remain true as stated.)

COROLLARY 5. If Γ is increasing in dominance and u is nondecreasing, then U is determined by the values that the gambles assign to the intervals (as are U_n and U_ω).

(Though Corollary 5 extends Corollary 4.5.1 from casinos to houses that increase in dominance, the corresponding extension of Theorem 4.5.2 is false. We have not determined whether Theorems 2 and 3 and Corollaries 4 and 5 apply to V.)

One particularly strong way for a house to increase in dominance is to have $\Gamma(f) \supset \Gamma(f')$ whenever $f \geq f'$; such a house *increases in freedom.*

LEMMA 1. If Γ increases in freedom and $f \geq f'$, then every strategy available in Γ at f' is also available at f.

THEOREM 4. If Γ increases in freedom, then V and U are nondecreasing for every u.

Suppose, to consider the most important sort of utility function, that, for some fortune g, $u(f) = 0$ for $f < g$ and $u(f) = 1$ for $f \geq g$. Just what functions U can occur for a Γ that increases in dominance? First, U must be nondecreasing according to Theorem 3. Second, for leavable Γ, U must be 1 at $f \geq g$. Third, either $\inf U = 0$ or $U \equiv 1$, as Theorem 3.8.5 says.

The three conditions are not only necessary; they are also sufficient, and even sufficient for a function U to be the U of the full house under U, which is of course increasing in freedom as well as in dominance. (Compare Theorems 7.4.2 and 7.4.3, and Examples 7.4.1 and 7.4.2.) Presumably, the situation with regard to V is not very different.

3 MONOTONE HOUSES

In this section the space of fortunes will be an interval unbounded from above, usually $(-\infty, \infty)$, $[0, \infty)$, or $(0, \infty)$; and in contrast to the preceding section, the arithmetic, or group theoretic, relations among the fortunes will play a role.

The set of lotteries available in Γ at f is $\Theta(f) = [\Gamma(f) - f]$ (according to the usage established for casinos in Section 4.1). Say that Γ is *increasing (not necessarily strictly) in displacement* if, for all f and f' with $f > f'$, $\Theta(f) \supset \Theta(f')$; it is *decreasing in displacement* if $\Theta(f) \subset \Theta(f')$ and *stationary (in displacement)* if $\Theta(f) = \Theta(f')$. In terms of $\Gamma(f)$, these concepts can also obviously be expressed thus: Γ is increasing, decreasing, or stationary in displacement if and only if, for all f and all positive t, $[\Gamma(f) + t]$ is a subset of, is a superset of, or is equal to $\Gamma(f + t)$. Since houses that are increasing in displacement are also increasing in dominance, conclusions in the preceding section apply to them.

There might, of course, be but one lottery θ available in the entire house. Such one-lottery, or random-walk, houses are stationary and correspond to sums of independent, identically distributed random variables. They will receive special attention in Chapter 10.

Henceforth, when F is $(-\infty, \infty)$, $u(f)$ will be understood to be 0 for negative f and 1 for nonnegative f. In this framework one might ask what functions U actually arise if Γ is a *monotone* house, that is, one which is increasing or decreasing in displacement.

THEOREM 1. These three conditions are equivalent for a U on $(-\infty, \infty)$:

(*a*) U is the U of a house that increases in displacement.

(*b*) U is the U of a house that increases in dominance.

(*c*) $0 \leq U \leq 1$; $U(f) = 1$ for $f \geq 0$; U is nondecreasing; and the infimum of U is 0, or $U \equiv 1$.

Proof: A house that increases in displacement certainly increases in dominance; so (*a*) implies (*b*), which in turn implies (*c*), as Theorems 2.3 and 3.8.5 make evident. It is also evident that 1 is the U of a house that increases in displacement. What remains is to see that (*c*) implies (*a*) when $\inf U = 0$.

Let $\Theta^*(g)$ consist of all one-point and two-point—or even of all—

lotteries available at g in the full house under U, and let $\Theta(f)$ be the intersection of all $\Theta^*(g)$ for $g \geq f$. Evidently, the gambling house thus constructed is leavable, increasing in displacement, and such that U is excessive for Γ.

Finally, as will now be shown, for each f and each positive ϵ less than $U(f)$, there is a two-point θ in $\Theta(f)$ for which $\theta\{-f\} = U(f) - \epsilon$. This is trivial if $U(f)$ is 0 or 1; otherwise, let $\theta\{e\} = 1 - (U(f) - \epsilon)$ for some e with $U(e) \leq \epsilon$. This $\theta \in \Theta(f)$, as is verified by calculating thus for $g \geq f$.

$$\begin{aligned}[\theta + g]U &= (U(f) - \epsilon)U(g - f) + (1 - (U(f) - \epsilon))U(g + e) \\ &\leq (U(f) - \epsilon)1 + 1\epsilon = U(f) \leq U(g). \quad \blacklozenge\end{aligned}$$

If a house on $(-\infty, \infty)$ is decreasing in displacement, a poor man can imitate a rich one; so a gambler whose fortune is $f + g$, where f and g are negative, can rise to at least f with probability nearly $U(g)$ and, if successful, continue on to 1 with probability at least $U(f)$, as Sections 2.9 and 2.10 make rigorous. Therefore,

$$U(f + g) \geq U(f)U(g) \qquad \text{for } f, g \leq 0. \tag{1}$$

Also, as usual, $U(f) = 1$ for $f \geq 0$, $0 \leq U \leq 1$; and $\inf U = 0$, or $U \equiv 1$.

Whether the conditions are sufficient for a U to be the U of a house decreasing in displacement we do not know. They obviously are if, and only if, U is the U of the largest subhouse of the full house under U that is decreasing in displacement; that is, $\Theta(f)$ is the intersection of $\Theta^*(g)$ for all $g \leq f$, where $\Theta^*(g)$ is, as in the proof of Theorem 1, the set of all lotteries θ for which $\int U(g + x)\, d\theta(x) \leq U(g)$. The U of this house is evidently majorized by U. If U approaches 0 as f approaches $-\infty$, the conditions are sufficient. For then U is attainable in Γ. In fact, for $f < 0$, let $\theta(f)\{-f\} = U(f)$ and $\theta(f)(-\infty, g) = 1 - U(f)$ for all $g < 0$. Then, for $g \leq f < 0$,

$$[\theta(f) + g]U = U(g - f)U(f) \leq U(g).$$

It would be interesting to know whether U is attainable if only gambles countably additive on the intervals are used. Even if this is not true in general, it might be under the conditions of the next theorem.

THEOREM 2. U on $(-\infty, \infty)$ is the U of a stationary house if and only if it satisfies (1) and condition (*c*) of Theorem 1.

Proof: Proceeds along the lines of the above discussion and the proof of Theorem 1. ◆

Whether V can replace U in (1) we do not know, but easy variants of (1) are

$$U_{m+n}(f + g) \geq U_m(f)U_n(g); \tag{2}$$

$$U_\omega(f + g) \geq U_\omega(f)U_\omega(g). \tag{3}$$

The U of a stationary house need be neither continuous nor strictly increasing.

Example 1. If $\Theta(f)$ consists, for every f, of the two-point lottery that attaches probability w to $+1$ and $\bar{w}$ to -1 for some w, $0 < w < \frac{1}{2}$, then $U(f)$ is well known and easily seen (by Theorem 2.12.1) to be $(\bar{w}/w)^{\langle f \rangle}$ for nonpositive f, where $\langle f \rangle$ is the greatest integer not greater than f.

If the U of a stationary house is continuous at 0, then it is continuous at all f, as follows immediately from (1).

Example 1 shows that the U of a stationary house need not be strictly increasing. Nonetheless, if $U(f) = 0$ for some negative f, then it is 0 for all f, as follows from (1) and the monotony of U.

We do not know whether a nonconstant U must be strictly increasing if it is continuous. Continuity and strict monotony of U are easy to establish for many stationary houses and some others, thus:

THEOREM 3. If, for some θ, $[t\theta]$ is available at every f in a Γ on $(-\infty, \infty)$ for all t in $(0, 1)$, $\theta[x, \infty) > 0$ for some positive x, and $\theta[-\infty, y)$ approaches 0 as y approaches $-\infty$, then U is continuous. If, in addition, $\theta(-\infty, y) = 0$ for some real y and $U(f) < 1$ for some f, then U is strictly increasing on $(-\infty, 0]$.

Proof: Apply Theorem 2.14.1. ◆

When the space of fortunes consists of the nonnegative real numbers, houses that are stationary or decreasing in displacement are of academic interest at most. For the lotteries in such houses—all of them contained in $\Theta(0)$—must be carried on the nonnegative real numbers; and gambling with no possible loss of money is not exciting.

The group of *dilatations* (*about* 0), the mappings that carry x into tx for positive t, plays the same role on $(0, \infty)$ as translations do on $(-\infty, \infty)$; for the positive numbers under multiplication are, by taking logarithms, isomorphic to the additive group of real numbers. A house Γ on $[0, \infty)$ or $(0, \infty)$ is *increasing, decreasing,* or *stationary in dilatation* according as $\Theta(tf) \supset [t\Theta(f)]$ for all $f > 0$ and for t in $[1, \infty)$, $(0, 1]$, or $(0, \infty)$. These relations can be expressed in just the same way in terms of $\Gamma(f)$; for $\Gamma(tf) = [\Theta(tf) + tf]$, and this includes $[t\Gamma(f)]$, or $[t\Theta(f) + tf]$, if and only if $\Theta(tf)$ includes $[t\Theta(f)]$.

In these terms, a casino is seen to be a gambling house on the nonnegative reals that is increasing in displacement and decreasing in dilatation, and in which $\Theta(0)$ contains only $\delta(0)$. A rich man's casino is one that is stationary in dilatation.

The problems of which functions can be the U of a house on $(0, \infty)$ that increases, decreases, or is stationary in dilatation when u is 0 for f in $(0, 1)$ and 1 in $[1, \infty)$ are isomorphic to the corresponding problems about displacement; so there is no need to paraphrase the preceding paragraphs of this section. The analogue of (1) is

$$U(fg) \geq U(f)U(g), \tag{4}$$

and the analogue of the exponential function with nonnegative λ is f^{λ}. The corresponding problems for U on $[0, \infty)$ seem uninteresting.

4. INCREASING FULL HOUSES

If the full house under v on an interval I, unbounded from above, increases in displacement, then v *increases in displacement.* The problem of the present section is to characterize such functions. A partial answer flows with but little effort from Theorem 7.3.2. Namely, v increases in displacement on $(-\infty, \infty)$ if and only if, for every f and g

with $g > f$, there is an $a(f, g) \geq 0$ such that

$$(1) \qquad v(g + \Delta) - v(g) \leq a(f, g)[v(f + \Delta) - v(f)]$$

for all Δ.

The extension of this partial answer to other intervals and, still more, the exploration of (1) cost some effort, which we hope will be repaid by its analytic if not its gambling interest.

The problem of this section bears on the theory of utility. For if v is interpreted as a person's utility function, then v increases in displacement if and only if the person's timidity increases with his wealth.

The function f^2 on $[0, \infty)$ is easily seen to be increasing in displacement, but there is no nonnegative $a(f, g)$ satisfying (1) for $f = 0$. Therefore, in extending the partial answer to an I with a lower endpoint, some special provision will have to be made. Also, it is sometimes useful to adapt (1) to all intervals open on the right. A function v on an interval J of positive length without an upper endpoint is *miserly* if for every f in J—except possibly if f is both the lower endpoint of J and a minimum of v—there is, for every $g > f$, an $a(f, g) \geq 0$ satisfying (1) for all Δ for which $(f + \Delta)$ and $(g + \Delta)$ are in J.

Of course, if v is miserly on J, so is $av + b$ for $a > 0$. Similarly, $v(af + b)$ is miserly on $(J - b)/a$. These trivial transformations will be freely used. When we say that f is an endpoint of J, we always mean, in part, that $f \in J$.

LEMMA 1. Suppose v on J is miserly and attains its minimum, say 0, at some g'. If g' is interior to J, then $v(f) = 0$ for every f interior to J. If g' is the lower endpoint of J, then v is nondecreasing on J and continuous from the right in the interior of J.

Proof: If g' is interior to J, $v(g') = 0$, $\Delta > 0$, and $(g' + \Delta)$ and $(g' - \Delta)$ are in J, then

$$(2) \quad \begin{aligned} v(g' + \Delta) = v(g' + \Delta) - v(g') &\leq a(g' - \Delta, g')[v(g') - v(g' - \Delta)] \\ &= -a(g' - \Delta, g')v(g' - \Delta) \leq 0. \end{aligned}$$

Therefore $v(g) = 0$ for every g greater than g'. Unless $v(g) = 0$ for every g interior to J, there is an f_0 interior to J such that $v(f) > 0$

for $f < f_0$ and $v(g) = 0$ for $g > f_0$. This implies a contradiction. For if $g - \Delta < f_0 < g$, then (1) implies that $a(g - \Delta, g) = 0$, but if $g - 2\Delta$ is also in J,

$$(3) \qquad 0 < v(g - \Delta) = v(g - \Delta) - v(g) \\ \leq a(g - \Delta, g)[v(g - 2\Delta) - v(g - \Delta)].$$

Suppose now that $0 \in J$ and is the lower endpoint of J, and $v(0) = 0$. Then

$$(4) \qquad v(g - f) - v(g) \leq a(f, g)[v(0) - v(f)] \leq 0$$

whenever $0 < f < g$; so $v(g)$ is nondecreasing. If, further, f is a point of continuity of v and Δ is small and positive,

$$(5) \qquad 0 \leq v(g + \Delta) - v(g) \leq a(f, g)[v(f + \Delta) - v(f)] \\ = a(f, g)o(1) = o(1). \quad \blacklozenge$$

THEOREM 1. A function v, on an interval I unbounded from above, increases in displacement if and only if it is miserly.

Proof: The "if" clause is evident when the exceptional possibility does not obtain. It follows in general with the help of Lemma 1. As Theorem 7.3.2 shows, the "only if" clause applies for all $g > f$, except possibly if $v(f) = \min v$ and f is not a lower endpoint of I. In the exceptional case, the penultimate sentence of Lemma 1 shows that v is constant in the interior of I; so (1) applies with $a(f, g) = 0$. ◆

For some insight into the implications of Theorem 1, suppose that v is rather regular.

LEMMA 2. If v on an open interval J has a finite derivative $\dot{v}$ that is positive (negative), then v is miserly if, and only if, $\log \dot{v}$ is concave ($\log -\dot{v}$ is convex).

Proof: The two cases are very similar; only that of positive $\dot{v}$ will be treated here. If v is miserly, f is interior to J, and $f < g$, then

$$(6) \qquad \dot{v}(g)\Delta + o(\Delta) \leq a(f, g)[\dot{v}(f)\Delta + o(\Delta)]$$

for small Δ; so

$$a(f, g) = \frac{\dot{v}(g)}{\dot{v}(f)}. \tag{7}$$

Applying (1) to f, g, and Δ for $\Delta > 0$ and to $f' = f + \Delta$, $g' = g + \Delta$, and $\Delta' = -\Delta$,

$$\begin{aligned} a(f, g)[v(f + \Delta) - v(f)] &\geq v(g + \Delta) - v(g) \\ &\geq a(f + \Delta, g + \Delta)[v(f + \Delta) - v(f)] \\ &> 0. \end{aligned} \tag{8}$$

In view of (7) and (8), $\dot{v}(g)/\dot{v}(f)$ is nonincreasing, and so $\log \dot{v}$ is concave.
Conversely, if $\log \dot{v}$ is concave, the mean value theorem shows easily that the a of (7) is effective in (1); so v is miserly. ◆

If a nonnegative function is concave, so is its logarithm; if the logarithm of a nonnegative function is convex, so is the function.

Example 1. The function $\lambda(\lambda - 1)f^\lambda$ on $(0, \infty)$ is miserly—or, equivalently, increasing in displacement—for all real λ. For $\lambda \geq 0$, $(\lambda - 1)f^\lambda$ is miserly on $[0, \infty)$, but, for $\lambda < 0$, with no finite value of $v(0)$ can f^λ be extended to be miserly on $[0, \infty)$.

Some functions increasing in displacement are more or less irregular.

Example 2. If $v(0)$ is arbitrary and $v(f) = 0$ for $f > 0$, then v is increasing in displacement on $[0, \infty)$.

Example 3. If v is nondecreasing and miserly on J, then so is $w = \min(v, 1)$.

Example 4. If v is nonincreasing and miserly on $[0, \infty)$, $v(0)$ can be arbitrarily increased without disturbing either property.

Example 5. If $H(f)$ is a Hamel function—that is, a not necessarily linear function on $(-\infty, \infty)$ for which $H(f + g) = H(f) + H(g)$—then $H(f)$ and $e^{H(f)}$ are increasing in displacement.

Example 5 shows that there are nonmeasurable functions that increase in displacement. Such irregular functions do not serve our present purposes; we want ultimately to confine the discussion to monotonic functions, but some readers will share our interest in seeing that a measurability assumption has the same effect. The rest of this section is only about such economy-of-regularity hypotheses.

LEMMA 3. If v is miserly on J, and f, g, $f + \Delta$, and $g + \Delta$ are in J with $g > f$, then $v(f + \Delta) - v(f) \leq 0$ (≥ 0) implies $v(g + \Delta) - v(g) \leq 0$ (≥ 0).

Proof: If Δ is nonnegative, the conclusion follows from (1), with one exception covered by the final sentence of Lemma 1. If Δ is negative, the same argument applies with $f' = f + \Delta$, $g' = g + \Delta$, and $\Delta' = -\Delta$ in place of f, g, Δ. ◆

(P. R. Halmos contributed to the proof of the next lemma.)

LEMMA 4. If v on $[0, 1)$ is Lebesgue measurable and miserly, then v is monotonic.

Proof: Let $P = \{\Delta\colon v(\Delta) \geq v(0)\}$ and $N = \{\Delta\colon v(\Delta) \leq v(0)\}$. In view of Lemma 3, if $0 \leq f \leq f + \Delta \leq 1$, then $v(f + \Delta) - v(f) \geq 0$ if $\Delta \in P$ and ≤ 0 if $\Delta \in N$. Therefore $P/n \subset P$ and $N/n \subset N$ for every positive integer n. Also, $(P/2) + (P/2) = P$, and $(N/2) + (N/2) = N$. Evidently P and N are Lebesgue measurable, and $[0, 1) = P \cup N$. So $(P/2) \cup (N/2) = [0, 1/2)$, and either $P/2$ or $N/2$ has positive measure. Therefore, $P = (P/2) + (P/2)$ or $N = (N/2) + (N/2)$ must have an interior point, according to a slight variant of a well-known fact (Halmos 1950, Theorem 16B, p. 68). It is now clear that P or N is $[0, 1)$. ◆

LEMMA 5. Let v be miserly on J. If v is monotonic on some subinterval K of positive length, then v is monotonic throughout J. If v is constant on K but is not constant on J with its lower endpoint deleted, then v is nondecreasing and, for some f_0 interior to J, v is strictly increasing in $J \cap (-\infty, f_0]$ and constant in $J \cap [f_0, \infty)$.

Proof: If v is not constant on K, the first assertion is easily derived from Lemma 3.

If v is constant, say α, on K, then it may be constant on the whole of J or on J minus its lower endpoint. In these trivial cases, herewith set aside, the lemma is obvious. Let f_0 be the least element of J such that $v(f) = \alpha$ for all $\alpha > f_0$; in the nontrivial cases f_0 exists and is interior to J, as Lemma 3 implies.

If v is monotonic in any interval of positive length to the left of f_0, it cannot be constant in that interval; so the first assertion of the lemma applies to show that v is strictly monotonic to the left of f_0. Otherwise, there exist f and g in J with $f < f_0 < g$ and $v(f) \neq \alpha$. Also, for some positive Δ, $g - \Delta > f_0$ and $v(f + \Delta) - v(f)$ and $\alpha - v(f)$ are of strictly opposite signs. For some positive integer n, $f_0 < f + n\Delta < g$, and $v(f + n\Delta) - v(f) = \alpha - v(f)$ is of the same sign as $v(f + \Delta) - v(f)$, which is a contradiction.

What remains is only to show that v on $J \cap (-\infty, f_0]$ does not exceed v on $J \cap [f_0, \infty)$. Consider f and g with $f < f_0 < g$ and $(2g - f)$ and $(2f - g)$ in J. Since

$$0 = v(2g - f) - v(g) \leq a(f, g)[v(g) - v(f)], \tag{9}$$

$v(g) \geq v(f)$ or $a(f, g) = 0$ and $v(g) < v(f)$. But the latter case leads to a contradiction:

$$\begin{aligned} 0 < v(f) - v(g) &= v(g - (g - f)) - v(g) \\ &\leq a(f, g)[v(2f - g) - v(f)] \\ &= 0. \quad \blacklozenge \end{aligned} \tag{10}$$

LEMMA 6. If v is miserly on J and is Lebesgue measurable in some subinterval of positive length, then v is monotonic on J.

Lemma 6 is reminiscent of, and in effect an extension of, the fact that a measurable Hamel function is linear. For, clearly, Hamel functions are miserly, and monotonic Hamel functions are linear.

LEMMA 7. If v is miserly and strictly monotonic on an open interval J, then, for all f in J, the derivative $\dot{v}(f)$ exists, is finite, and is different from 0.

Proof: Consider f, g, and h in J, with $f < g < h$, and such that $\dot{v}(f)$ and $\dot{v}(h)$ exist and are finite. Since h is not a maximum for v, $a(g, h) > 0$.

$$\begin{aligned} a(f, g)[v(f + \Delta) - v(f)] &= a(f, g)\dot{v}(f)\Delta + o(\Delta) \\ &\geq v(g + \Delta) - v(g) \\ &\geq \frac{1}{a(g, h)} [v(h + \Delta) - v(h)] \\ &= \frac{\dot{v}(h)}{a(g, h)} \Delta + o(\Delta). \end{aligned} \tag{11}$$

Therefore,

$$a(f, g)\dot{v}(f) = \frac{\dot{v}(h)}{a(g, h)} = \dot{v}(g). \tag{12}$$

So $\dot{v}(g)$ exists and is finite for all g in J. Since v is therefore absolutely continuous, it can now be assumed that $\dot{v}(h) \neq 0$ and concluded that $\dot{v}(g) \neq 0$. ◆

LEMMA 8. If v on $[0, 1)$ is miserly and nondecreasing and if v is not constant on $(0, 1)$, then v is continuous on $[0, 1)$.

Proof: In view of Lemmas 5 and 6, the only possible discontinuities are at 0 and at the f_0 of Lemma 5. Also, according to Lemma 5, v is strictly monotonic in $(0, 1)$ or in $(0, f_0)$.

Consider first the behavior of v at 0. For fixed Δ and small positive ϵ,

$$\begin{aligned} v(\Delta) - v(\Delta + \epsilon) &\leq a(\epsilon, \Delta + \epsilon)[v(0) - v(\epsilon)] \\ &= \frac{\dot{v}(\Delta + \epsilon)}{\dot{v}(\epsilon)} [v(0) - v(\epsilon)] \end{aligned} \tag{13}$$

in view of (12). Since v is continuous at Δ and increasing at 0 and since, in view of Lemma 2, $\dot{v}(\Delta + \epsilon)/\dot{v}(\epsilon)$ is nonincreasing in ϵ, $v(\epsilon)$ must approach $v(0)$.

According to Lemma 1, v is continuous to the right for $f > 0$; so the only remaining possibility of a discontinuity is a jump down from f_0. For $0 < \epsilon < \epsilon_0 < f_0 - \Delta$,

$$\begin{aligned}(14) \qquad 0 &< v(f_0) - v(f_0 - \epsilon) \\ &\leq \frac{\dot{v}(f_0 - \epsilon)}{\dot{v}(f_0 - \epsilon - \Delta)} [v(f_0 - \Delta) - v(f_0 - \Delta - \epsilon)] \\ &\leq \frac{\dot{v}(f_0 - \epsilon_0)}{\dot{v}(f_0 - \epsilon_0 - \Delta)} O(\epsilon). \quad \blacklozenge\end{aligned}$$

THEOREM 2. If v on J is Lebesgue measurable on a subinterval of positive length, then v is miserly if and only if it is of one of the following types:

(*a*) v is constant, except possibly at the lower endpoint of J.

(*b*) v has an everywhere finite derivative $\dot{v}$, which is positive except possibly at the lower endpoint of J, and $\log \dot{v}$ is concave.

(*b'*) v is continuous, and, for some f_0 interior to J, it is of type (*b*) in $J \cap (-\infty, f_0]$ and constant in $J \cap [f_0, \infty)$.

(*c*) v has an everywhere negative derivative $\dot{v}$, which is finite except possibly at the lower endpoint of J, and $\log -\dot{v}$ is convex.

(*c'*) v is strictly decreasing on J and of type (*c*) on J minus its left endpoint.

COROLLARY 1. If v and $-v$ are both measurable and miserly, then v is of type (*a*), linear, or of the form $ae^{f\lambda} + b$.

THEOREM 3. If w is such that $w(v)$ is miserly whenever v is, then w is linear and nondecreasing; if $w(v)$ is miserly whenever v is miserly and nondecreasing, then w is of the form $\min(af + b, c)$ for some $a \geq 0$.

5. FULL-HOUSE CASINOES

This section studies under what conditions a full house under a function v is a casino. The remaining two sections of the chapter do not depend on this one, but there are a few references to this section in the next chapter.

What to require of $\Gamma(0)$ in a casino and whether to regard 0 as an element of a casino at all are petty matters. Until now it has been most convenient to include 0 and to require $\Gamma(0)$ to be $\{\delta(0)\}$, but full houses that are casinos in just that sense are rare and uninteresting. A slightly modified definition is therefore adopted here: A gambling house Γ on $[0, \infty)$ or on $(0, \infty)$ is a *casinoe*—rhymes with "canoe"—if it is leavable, if $\Theta(g) \supset \Theta(f)$ for $g > f$ (increasing in displacement), and if $\Theta(tf) \supset [t\Theta(f)]$ for $0 < f$ and $0 < t < 1$ (decreasing in dilatation relative to $(0, \infty)$).

THEOREM 1. If Γ is a casino, it is also a casinoe on $[0, \infty)$. If Γ' is a casinoe on either $[0, \infty)$ or $(0, \infty)$, and $\Gamma(f) = \Gamma'(f)$ for $f > 0$ and $\Gamma(0) = \{\delta(0)\}$, then Γ is a casino, and $U(f) = U'(f)$ for $f > 0$.

COROLLARY 1. On $[0, \infty)$, U' is the U of some casinoe if and only if U' is the U of a casino or $U' \equiv 1$. On $(0, \infty)$, U' is the U of a casinoe if and only if U' coincides on $(0, \infty)$ with the U of a casino.

If the full house Γ_v under v on $[0, \infty)$ or on $(0, \infty)$ is a casinoe, then v is a *casinoe maker*. A preliminary characterization of casinoe makers on $(0, \infty)$ is almost immediate from Theorem 4.1 and the isomorphism $f' = -\log f$. In view of this isomorphism, $v(f)$ on $(0, \infty)$ decreases in dilatation if and only if $v(e^z)$ on $(-\infty, \infty)$ decreases in displacement.

LEMMA 1. On $(0, \infty)$, v is a casinoe maker if and only if, for all f and g with $0 < f < g$, there are nonnegative $a(f, g)$ and $b(f, g)$ such that

$$v(g + \Delta) - v(g) \leq a(f, g)[v(f + \Delta) - v(f)] \tag{1}$$

and

$$b(f, g)[v(g\lambda) - v(g)] \geq v(f\lambda) - v(f) \tag{2}$$

for all $\Delta > -f$ and $\lambda > 0$.

The next lemma follows easily from the definition of casinoe and from consideration of the available one-point lotteries.

LEMMA 2. Each casinoe maker on $(0, \infty)$ is strictly monotonic or constant.

Lemma 2 is reminiscent of the fact that the only isomorphism of the (semi) field of (positive) real numbers into itself is the identity (Dieudonné 1960, p. 83).

LEMMA 3. On $(0, \infty)$, v is a casinoe maker if and only if v is differentiable and satisfies one of the following three conditions:

(*a*) $\dot{v}$ is strictly positive, $\log \dot{v}(f)$ is concave in f, and $\log \dot{v}(e^z)$ is convex in z.

(*b*) $\dot{v}$ is strictly negative, $\log -\dot{v}(f)$ is convex in f, and $\log -\dot{v}(e^z)$ is concave in z.

(*c*) v is constant.

Proof: Case (*c*) is obviously a possibility. Failing case (*c*), a casinoe maker is strictly increasing or strictly decreasing according to Lemma 2. The proof is easily completed by application of Lemma 1 and Theorems 4.1 and 4.2 to $v(f)$ and to $v(e^{-z})$. ◆

Example 1. (*a*) Let $v(f) = e^{\lambda f}$ for some $\lambda > 0$. Then $\log \dot{v}(f) = \lambda f + \log \lambda$ is concave, barely; and $\log \dot{v}(e^z) = \lambda e^z + \log \lambda$ is strictly convex.

(*b*) $v(f) = e^{-\lambda f}$ for some $\lambda > 0$.

Example 2. (*a*) Let $v(f) = f^\alpha$ for some $\alpha \geq 1$. Then $\log \dot{v}(f) = \log \alpha + (\alpha - 1) \log f$ is concave—strictly concave, unless $\alpha = 1$; and $\log \dot{v}(e^z) = \log \alpha + (\alpha - 1)z$ is barely convex.

(*b*) $v(f) = f^\alpha$ for some $\alpha < 0$.

(*b*′) Let $v(f) = -\log f$ to see a casinoe maker that is bounded neither from above nor from below.

(*b*″) $-f^\alpha$ for $0 < \alpha \leq 1$.

A casinoe maker is not only differentiable; it is twice differentiable and nearly three times differentiable.

THEOREM 2. The following conditions are equivalent on $(0, \infty)$:

(*a*) v is a nonconstant casinoe maker.

(*b*) v, $-\ddot{v}/\dot{v}$, and $f\ddot{v}/\dot{v}$ are all nondecreasing or all nonincreasing.

(*c*) $\ddot{v}$ is absolutely continuous, $\ddot{v} > 0$, and, introducing q for $\ddot{v}/\dot{v}$,

$$-1 \leq \frac{f\dot{q}}{q} \leq 0 \tag{3}$$

except for a set of Lebesgue measure 0, unless $\ddot{v} \equiv 0$.

Proof: Suppose v is as in case (a) of Lemma 3, and let F_R and F_L denote right and left differentiation with respect to f, and Z_R and Z_L with respect to z.

$$(4) \qquad F_R \log \dot{v}(f) = \frac{F_R\dot{v}(f)}{\dot{v}(f)} \leq \frac{F_L\dot{v}(f)}{\dot{v}(f)},$$

$$(5) \qquad Z_R \log \dot{v}(e^z) = \frac{Z_R\dot{v}(e^z)}{\dot{v}(e^z)} = e^z \frac{F_R\dot{v}(e^z)}{\dot{v}(e^z)} \geq e^z \frac{F_L\dot{v}(e^z)}{\dot{v}(e^z)}.$$

Therefore, $F_R\dot{v}(f) = F_L\dot{v}(f)$, and $\ddot{v}(f)$ exists. A similar argument applies to case (b) of Lemma 3. Conditions (a) and (b) of the present theorem are now seen to be equivalent, in the light of Lemma 3, by elementary calculation.

Now suppose condition (b) obtains. If q were 0 at one point, it would be identically 0. Since q must be positive or negative according as $\dot{v}$ is positive or negative, $\ddot{v}$ is always positive. If, for instance, q is positive and decreasing,

$$(6) \quad 0 \leq q(f) - q(f + \Delta) \leq \frac{f + \Delta}{f} q(f + \Delta) - q(f + \Delta)$$
$$= \frac{\Delta}{f} q(f + \Delta).$$

Since q is a bounded function of f for f bounded away from 0, (6) shows that q is absolutely continuous. Since $\ddot{v}$ exists everywhere and is positive, $\dot{v}$ is absolutely continuous. Now that $\ddot{v}$ is seen to be the product of two absolutely continuous functions, it, too, is absolutely continuous. As then follows easily by differentiating the terms in (b), $-1 \leq f\dot{q}/q \leq 0$ whenever $\dot{q}$ exists. Similarly, (c) implies (b). ♦

The characterization of casinoe makers on $(0, \infty)$ achieved by Theorem 2 is transferred to the more interesting interval $[0, \infty)$ by the next theorem, proof of which, being both straightforward and somewhat tedious, is omitted. Natural tools are Theorem 4.2, Lemma 1, the extension of (2) to $\lambda = 0$ for a casinoe maker on $[0, \infty)$, the uniqueness of $b(f, g)$ for strictly decreasing casinoe makers implied by the multiplicative form of (4.10), and Lemma 2.

THEOREM 3. On $[0, \infty)$, v is a casinoe maker if and only if one of the following conditions obtains.

(*a*) v is strictly increasing and continuous, and v restricted to $(0, \infty)$ is a casinoe maker there.

(*b*) v is strictly decreasing and continuous, and v restricted to $(0, \infty)$ is a casinoe maker there.

(*c*) v is constant on $(0, \infty)$.

THEOREM 4. The restriction of each casinoe maker on $[0, \infty)$ to $(0, \infty)$ is a casinoe maker on $(0, \infty)$. For strictly increasing casinoe makers on $[0, \infty)$, this mapping is one-to-one and onto the strictly increasing casinoe makers on $(0, \infty)$. For strictly decreasing casinoe makers on $[0, \infty)$, this mapping is also one-to-one, and its range is strictly decreasing casinoe makers on $(0, \infty)$ that are bounded from above.

Proof: Since increasing casinoe makers are increasing convex functions and so bounded from below, this theorem is immediate from Theorem 3. ♦

Part (*b*) of Example 1 shows that a decreasing casinoe maker on $(0, \infty)$ may not be extendible to $[0, \infty)$. In connection with the usual u, only strictly increasing casinoe makers are interesting. Attention is therefore confined now to strictly increasing casinoe makers on $[0, \infty)$. These, so normalized that $v(0) = 0$ and $v(1) = 1$, will be called *normalized casinoe makers.*

THEOREM 5. If v is a normalized casinoe maker, then v restricted to $[0, 1]$ coincides with the U of the casinoe that is the full house under v, and hence is a casino function.

Proof: See Example 7.4.1. ♦

Theorem 5 provides a fertile way to identify casino functions—though of a very limited sort—as the next paragraph illustrates.

Lawrence Jackson was the first to show that, for each α between 1 and 2, convex combinations of f and f^α are casino functions. His proof was a direct verification that such convex combinations satisfy the casino inequality (4.2.1). Here is an alternative proof which shows that such convex combinations are not only casino functions but normalized casinoe makers.

THEOREM 6. The full house under a convex combination of f and f^α for $1 < \alpha \leq 2$ is a casinoe.

Proof: Let $v(f)$ be a convex combination of f and f^α. Then $q(f) = \ddot{v}(f)/\dot{v}(f) = kf^{\alpha-2}/(1 + cf^{\alpha-1})$ for some positive k and c; $q(f)$ is decreasing in f, and $fv(f)$ is increasing in f. Apply Theorems 2, 3, and 5. ♦

Convex combinations of casinoe makers are not generally casinoe makers. For, as reported in Section 4.3, Lawrence Jackson and Roger Purves have shown that, for $\alpha > 2$, no proper convex function of f and f^α is a casino function, let alone a normalized casinoe maker.

The rest of this section is about the generation of new casinoe makers from old ones. In the first place, it is not easy to generate new ones by composition of old ones. This is important only insofar as it explains the interest of later theorems, and so we do not give the proof in detail.

THEOREM 7. The function $w(v)$ ($v(w)$) is a strictly increasing casinoe maker on $(0, \infty)$ whenever v itself is if and only if w is strictly increasing and linear (strictly increasing, linear, and positive for positive f).

Proof: It is easy to see by testing with $v(f) = f$ that w must be a strictly increasing casinoe maker. Exploration of the test functions $v(f) = a + be^{\lambda f}$ for $\lambda > 0$ shows that this is impossible except for linear w. ♦

There is some freedom of composition for the derivatives of casinoe makers.

THEOREM 8. $y(x)$ is the derivative of a strictly increasing casinoe maker on $(0, \infty)$ whenever x is if and only if y is of the form af^c for some $a > 0$ and $c \geq 0$.

Proof: According to Theorem 2, what is required is that y be positive and that $-\dot{y}(x)\dot{x}/y(x)$ and $f\dot{y}(x)\dot{x}/y(x)$ be nondecreasing in f, whenever x is a positive function for which $-\dot{x}/x$ and $f\dot{x}/x$ are nondecreasing. Testing with $x(f) = f$, you see that y is positive and $-\dot{y}/y$ and $f\dot{y}/y$ are nondecreasing. Testing with $x(f) = \alpha e^f$, you see that

$x\dot{y}(x)/y(x)$ is nonincreasing in x and therefore constant. The only possibilities remaining for y are now af^c, with $a > 0$ and $c \geq 0$, and with such a y, $\dot{y}(x)\dot{x}/y(x) = c\dot{x}/x$. ◆

The question about $x(y)$ is more complicated. We merely present a not very satisfying theorem and give an example.

THEOREM 9. $x(y)$ is the derivative of a strictly increasing casinoe maker on $(0, \infty)$ whenever x is if and only if y is positive on $(0, \infty)$, and $-\dot{y}(f)$ and $f\dot{y}(f)/y(f)$ are nondecreasing.

Proof: What is required of y is that $-\dot{x}(y)\dot{y}/x(y)$ and $f\dot{x}(y)\dot{y}/x(y)$ be defined and nondecreasing whenever x is positive and $-\dot{x}/x$ and $f\dot{x}/x$ are nondecreasing. Of course, y has to be positive if $x(y)$ is to be defined. Testing with $x(f) = f$ shows that $f\dot{y}(f)/y(f)$ must be nondecreasing. Testing with $x(f) = e^f$ shows that $\dot{y}$ is nonincreasing.

The conditions are sufficient: To begin with, they imply that $\dot{y} \geq 0$; otherwise y would become negative for large f. Since $\dot{x}(y)/x(y)$ is nonnegative and nonincreasing in y, $\dot{x}(y)\dot{y}/x(y)$ is now seen to be nonincreasing in f. Similarly, $f\dot{x}(y)\dot{y}/x(y) = (f\dot{y}/y)(y\dot{x}(y)/x(y))$ is nondecreasing in f. ◆

Example 3. If $y(f) = f^c$, then $-\dot{y}(f) = -cf^{c-1}$ and $f\dot{y}(f)/y(f) = c$ are nondecreasing if and only if $0 \leq c \leq 1$.

Associated with a strictly increasing nonlinear casinoe maker v on $(0, \infty)$ is the function B for which

$$B(f) = \frac{\dot{v}(f)}{\ddot{v}(f)}. \tag{7}$$

Theorem 2 shows that B is positive and that B and $-B/f$ are nondecreasing. Call such a function a *casino basement* and call B *the basement* of v.

THEOREM 10. Every casino basement is the basement of exactly one normalized casinoe maker.

Example 4. Examples 1(a) and 2(a) have the basements λ^{-1} with $\lambda > 0$ and f/c with $c > 0$.

If (7) looks upside down, that is mainly to ensure that the identity function be a basement, which makes the next theorem prettier than it would otherwise be.

THEOREM 11. The function $C(B)$ $(B(C))$ is a casino basement whenever B is if and only if C is a casino basement.

Proof: Testing with $B(f) = f$ shows that each condition requires C to be a casino basement. If B and C are both casino basements, plainly $C(B)$ is nondecreasing. Similarly,

$$\frac{C(B(f))}{f} = \frac{C(B(f))}{B(f)} \frac{B(f)}{f}$$

is nonincreasing. ◆

COROLLARY 2. Casino basements form a semigroup with respect to composition.

Example 5. The function af^b is a casino basement if and only if $a > 0$ and $0 \leq b \leq 1$. These functions obviously constitute a noncommutative two-parameter semigroup; $(a', b') \cdot (a, b) = (a'a^{b'}, b'b)$. If v has af^b as its basement, then

$$\dot{v}(f) = K \exp \frac{f^{1-b}}{a(1-b)} \tag{8}$$

for some positive K.

Example 6. Let p, q, r, and s be real numbers; then

$$\frac{pf + q}{rf + s}$$

is a casino basement if and only if it is constant or no two of the coefficients p, q, r, and s have opposite signs and the determinant $d = (ps - qr)$ is positive. For the casinoe makers corresponding to the three-parameter semigroup with $d > 0$,

$$\dot{v}(f) = \begin{cases} K\left(f + \dfrac{q}{p}\right)^{d/p^2} e^{rf/p} & \text{for } p \neq 0, \\ Ke^{f(rf+2s)/2q} & \text{for } p = 0. \end{cases} \tag{9}$$

The first part of Theorem 11 has an easy extension to functions ϕ of n variables defined on $(0, \infty)^n$.

THEOREM 12. $\phi(B_1, \cdots, B_n)$ is a casino basement whenever $B_1, \cdots, B_n$ are casino basements if and only if ϕ is positive, is nondecreasing in each argument, and $\phi(tx_1, \cdots, tx_n) \leq t\phi(x_1, \cdots, x_n)$ for all positive $x_1, \cdots, x_n$ and for $t > 1$.

Example 7. Any function ϕ of n arguments that is positive, nondecreasing in each argument, and homogeneous of degree 1—in short, any function that could be called a "mean"—satisfies the condition of Theorem 12. In particular, let $\alpha_i \geq 0$, for $i = 1, \cdots, n$, and $\Sigma\, \alpha_i = 1$, and consider:

$$\sum (\alpha_i x_i^r)^{1/r} \qquad \text{for } r \neq 0, \tag{10}$$

$$\prod x_i^{\alpha_i}, \tag{11}$$

$$\max \alpha_i x_i, \tag{12}$$

$$\min \alpha_i x_i. \tag{13}$$

Example 7 shows that the class of casino basements is convex in the ordinary sense, and in many other senses as well.

As Theorem 2 makes clear, a casino basement B is absolutely continuous and leads to a *sub-basement* S defined by

$$S(f) = f\frac{d}{df}\log B(f) = \frac{f\dot{B}(f)}{B(f)} \tag{14}$$

almost everywhere. Modulo null functions, the sub-basements are simply the measurable functions on $(0, \infty)$ that are bounded between 0 and 1.

Strictly increasing, convex casinoe makers v and v' clearly have the same sub-basement if and only if B and B' are proportional to each other, that is, if and only if $\dot{v}' = a\dot{v}^c$ for some positive a and c. Compare Theorem 8.

Example 8. If $S \equiv b$ with $0 \leq b \leq 1$, then $B(f) = af^b$ for some $a > 0$. See Example 5.

Of course, the set of all sub-basements is convex in the ordinary and in practically any other sense, and its extreme points are in many senses the indicator functions of measurable sets.

6. STATIONARY FULL HOUSES

If the full house under v on $(-\infty, \infty)$ is stationary in displacement, then v is *stationary in displacement.* According to Theorem 4.1, v is stationary in displacement if and only if $v(f)$ and $v(-f)$ are both miserly. According to Corollary 4.1, if v is measurable throughout some interval of positive length, v is then stationary in displacement if and only if $\dot{v}$ is constant or $\log \dot{v}$ is linear or $\log -\dot{v}$ is linear. This conclusion will now be rederived in a more familiar way, with the slight advantage of finding the nonmeasurable possibilities.

The full house under a constant consists of all gambles and is therefore stationary in displacement. Setting this trivial possibility aside, suppose that v is a nonconstant function stationary in displacement. According to Lemma 4.1, it is impossible that v should attain a minimum. To postulate for the moment that $v(0) = 0$ and that 0 is neither a maximum nor a minimum of v is now a mere affine renormalization. Exploiting the fact that (4.1) must apply with $g = 0$ regardless of the sign of f (because v is miserly in both directions), you can easily verify (along the lines of the proof of part (c) of Theorem 7.3.3) that for some positive function a

$$v(f' + f) = a(f)v(f') + v(f) \tag{1}$$

for all f and f'. The problem remaining is to characterize the not identically vanishing solutions of (1) with $a(f)$ positive. This latter problem (with references to Nagumo, de Finetti, and Jessen) is essentially solved in (Hardy, Littlewood, and Polya 1934, pp. 68, 69) as follows. Equation (1) implies that $a(0) = 1$ and also $a(f)v(f') = a(f')v(f) + v(f')$, whence $a(f) = kv(f) + 1$ for some constant k, and (1) is equivalent to

$$v(f' + f) = kv(f)v(f') + v(f') + v(f), \tag{2}$$

with the condition $kv(f) + 1 > 0$.

If $k = 0$,

$$v(f' + f) = v(f') + v(f), \tag{3}$$

or v is additive. In this case, as is well known, v is either simply of the form af or else extremely wild—nonmeasurable with a graph dense in the whole plane (Hardy, Littlewood, and Polya 1934, p. 96).

If $k \neq 0$, (2) is equivalent to

$$a(f' + f) = a(f)a(f'), \tag{4}$$

where $a(f) = kv(f) + 1$ and is strictly positive. Therefore $\log a$ exists and satisfies (3). In summary:

THEOREM 1. The full house under v is stationary in displacement if and only if v is of one of the following mutually exclusive types or an affine transformation thereof.

(*a*) $v(f) = e^{\lambda f}$ for some $\lambda \neq 0$.
(*b*) $v(f) = f$.
(*c*) $v(f)$ is a discontinuous solution of (3).
(*d*) $\log v(f)$ is a discontinuous solution of (3).
(*e*) $v(f) = 0$ for all f.

Of the five possibilities listed in Theorem 1, only (*a*), and perhaps (*b*), will ordinarily be of interest. Possibilities (*b*) and (*c*), being unbounded from below, are not of interest in connection with any bounded utility u. Possibilities (*c*) and (*d*) allow one-point transitions from each fortune to an everywhere dense set of fortunes and are therefore of no interest in connection with even moderately regular utilities; possibility (*e*) allows all gambles at each fortune and is, therefore, of no interest in connection with any utility.

Since the full house under v on the real line is increasing in displacement if and only if the full house under v^* on the positive reals is increasing in dilatation, where $v^*(f) = v(\log f)$, and since a casino is a rich man's casino if and only if it is stationary in dilatation, Theorem 1 has this corollary:

COROLLARY 1. The full house Γ under a function on $[0, \infty)$ is a fair or subfair rich man's casinoe if and only if Γ is the full house under f^λ for some $\lambda \geq 1$.

COROLLARY 2. The full house Γ under a function on $(0, \infty)$ is a superfair rich man's casinoe if and only if Γ is the full house under a constant or under one of the functions in part (b), (b'), or (b'') of Example 5.2.

Recognition that the characteristic feature of a poor man's casinoe on $(0, \infty)$ is an adaptation to $(0, \infty)$ of stationarity in displacement leads, with a little effort, to the conclusion that the only full houses on $(0, \infty)$ that are poor man's casinoes are the full houses under $e^{\lambda f}$ for some λ, or under f or $-f$.

7. EXPONENTIAL HOUSES

The importance of the λ-*exponential houses*, that is, the full houses under functions of the form $e^{\lambda f}$ is suggested by Theorem 6.1. In the λ-exponential house a gambler whose fortune is f may use any gamble γ for which $\int e^{\lambda g}\, d\gamma(g) \leq e^{\lambda f}$.

This section is largely a report of the research of others. For the topic of exponential houses has a long history, mainly under somewhat different guises. We hope, though, that there are some new points here and that this collation and unification of the material will be useful.

De Finetti (1939, 1940) introduced and studied the concept of insurance companies that sell only lotteries, for which the expected value of $e^{\lambda f}$ is 1 for some negative λ, evaluating the probability that such an insurance company will ever go bankrupt starting with a positive fortune. It is but a change in conventions and in mathematically irrelevant value judgments to view de Finetti's problem as that of a gambler who wants to arrive at a nonnegative fortune starting with a negative one in a house that conserves $e^{\lambda f}$ for some positive λ. De Finetti's work on exponential houses has greatly influenced this section. De Finetti traces his own ideas on the topic through (Bertrand 1907, Section 92) to De Moivre (1711, Problem IX; 1718, Problem VII).

The considerable literature about a topic known as Wald's fundamental identity of sequential analysis is related to houses that conserve $e^{\lambda f}$ for some λ. See, for example, (Wald 1947, pp. 159–160), (Doob 1953, pp. 350–352), and (Bahadur 1958) for various points of view on this identity and for further references. For still other references and for presentation of closely related matters see also (Kemperman 1961) and (Hoeffding 1963). Still others have touched on exponential houses, as later parts of this section will bring out.

The importance of exponential houses is suggested by Theorem 6.1, according to which the functions U_λ, where $U_\lambda(f) = e^{\lambda f}$, are (except for oriented affine transformations) the only functions that generate interesting full houses stationary in displacement.

Since

$$\begin{aligned}[\theta + f]U_\lambda &= \int U_\lambda(g + f)\, d\theta(g) \\ &= U_\lambda(f) \int U_\lambda(g)\, d\theta(g) \\ &= U_\lambda(f)(\theta U_\lambda),\end{aligned}$$

the gamble $[\theta + f]$ conserves, decreases, or increases U_λ at f according as θ conserves, decreases, or increases U_λ at 0. In particular, the gamble $[\theta + f]$ is available at f in the λ-exponential house if and only if $\int e^{\lambda g}\, d\theta(g) \leq 1$. For this reason, the *Laplace transform* L_θ (or simply L) of the lottery θ is of interest for exponential houses, where

$$L_\theta(\lambda) = \theta U_\lambda = \int e^{\lambda f}\, d\theta(f). \tag{1}$$

The following theorem is a list of facts about the Laplace transform. Except possibly for a few easy details, these are widely known and used; see, for example, (de Finetti 1939), (Wald 1947), and (Girshick and Savage 1951). Proofs of the less obvious assertions will be found in (Widder 1941, Sections VI1 and VI2). To conform with convention, and for simplicity, the theorem is confined to lotteries θ for which $\theta(-g, g)$ approaches 1 as g approaches ∞.

THEOREM 1. If L is the Laplace transform of a lottery θ for which $\theta(-g, g)$ approaches 1 as g approaches ∞, then:

(*a*) $L(0) = 1$.

(*b*) $0 < L(\lambda) \leq \infty$.

(*c*) $\{\lambda: L(\lambda) < \infty\}$ is an interval C. (Any interval that contains 0 can occur.)

(*d*) On C, $L(\lambda)$ is continuous and strictly convex, unless $\theta(-\epsilon, \epsilon) = 1$ for every positive ϵ, in which case $L(\lambda)$ is 1 for all λ.

(*e*) With the possible exception mentioned in (*d*), $L(\lambda) = 1$ for at most one λ different from 0, say $\lambda(\theta)$.

(*f*) $\{\lambda: L(\lambda) \leq 1\}$ is a closed interval D, and the possible configurations are these: (i) There is a $\lambda(\theta)$, and D is the interval joining 0 to $\lambda(\theta)$; (ii) D consists of 0 alone; (iii) $D = C \cap (-\infty, 0]$, or $D = C \cap [0, \infty)$.

(*g*) In the interior of C, L is analytic.

(*h*) In the interior of C, (1) can be differentiated any number of times under the integral sign. With usual conventions about infinite and about one-sided derivatives, this applies even to the endpoints of C if C does not consist of 0 alone. In particular,

$$\frac{d^n}{d\lambda^n} L(\lambda) \bigg|_{\lambda=0} = \int f^n \, d\theta(f), \tag{2}$$

provided C does not consist of 0 alone.

(*i*) If, for some positive ϵ, $\theta(\epsilon, \infty) > 0$, then $L(\lambda)$ is not bounded in $[0, \infty)$.

(*j*) If, for some positive ϵ, $\theta(\epsilon, \infty) > 0$, θ is subfair, and C includes $[0, \infty)$, then $\lambda(\theta)$ exists.

To see the small changes entailed in Theorem 1 by inclusion of lotteries for which $\theta(-g, g)$ is uniformly less than 1, consider first the Laplace transform L_+ of any lottery θ_+ concentrated in the neighborhood of $+\infty$; $L_+(\lambda) = 0$ for $\lambda < 0$, $= 1$ for $\lambda = 0$, $= +\infty$ for $\lambda > 0$. Similarly for the Laplace transform of a lottery concentrated near $-\infty$. The most general L_θ is a convex combination of L_+, L_-, and an L of the type covered by Theorem 1. If L_+ and L_- both have positive coefficients in this convex combination, then $L_\theta(\lambda) = \infty$ except when $\lambda = 0$. If, for example, L_- has a positive coefficient, then $L_\theta(\lambda) = \infty$ for $\lambda < 0$ and $= 1$ for $\lambda = 0$, and $L_\theta(0 + 0) < 1$ if L_θ is finite for some positive λ. Actually, such a jump down to the right or left of 0 is the only possible discontinuity for a Laplace transform, apart from the jump to ∞ from the endpoints of C. As is easily checked, Theorem 1 remains correct without the exclusion of any lotteries, except for the possibility of a jump down at 0 and the possibility that $L_\theta(\lambda)$ vanishes if θ is θ_+ or θ_-.

Most of the parts of Theorem 1 have direct implications for the theory of exponential houses. In particular, part (*e*) implies that only trivial lotteries can belong to the λ-conserving house for more than one nonzero λ, and part (*j*) gives a frequently applicable condition for the existence of one such λ, $\lambda(\theta)$. Again, part (*f*) implies that if, for example, $0 \leq \lambda < \lambda'$, then the λ'-exponential house is a subhouse of the λ-exponential house; this monotony is strict, since two-point λ-conserving gambles exist.

Since

$$(3) \qquad [t\theta]U_\lambda = \int e^{\lambda f}\, d[t\theta](f)$$

$$= \int e^{t\lambda f}\, d\theta(f) = \theta U_{t\lambda},$$

$[t\theta]$ conserves U_λ at 0 if and only if θ conserves $U_{t\lambda}$ at 0. In view of part (*f*), if θ depresses U_λ, so does $[t\theta]$ for t in $[0, 1]$.

Let u be the indicator function of the nonnegative numbers. With this u, what is U for the λ-exponential house? According to Section 7.4, $U \equiv 1$ if $\lambda \leq 0$, and $U = \min(U_\lambda, 1)$ for positive λ. This evaluation applies to any subhouse of the λ-exponential house that contains all two-point, λ-conserving gambles. It is of primary interest for positive λ, and λ will be assumed positive throughout much of the rest of the section.

For any subhouse of the λ-exponential house ($\lambda > 0$), $U_\lambda(f) = e^{\lambda f}$ is an upper bound for $U(f)$; if a gambler is restricted to lotteries θ for which $\int e^{\lambda g}\, d\theta(g) \leq 1$, then the probability that his fortune ever increases by x is bounded by $e^{-\lambda x}$ for each $x > 0$. In applications, this upper bound is often close, as has been emphasized by de Finetti (1939) and by Wald (1947). Suppose, for example, that each $\Theta(f)$ contains a certain nontrivial θ that conserves $e^{\lambda f}$, $\lambda > 0$. Study the strategy σ initiating at f, $f < 0$, and repeating θ indefinitely or until f_n is first nonnegative. The following martingale-like formula can be interpreted and justified even in a finitely additive theory.

$$e^{\lambda f} = E_\sigma(e^{\lambda f_t}) = E_\sigma(e^{\lambda f_t}|f_t \geq 0)u(\sigma) \geq u(\sigma). \tag{4}$$

Any upper bound on the conditional expectation in (4) amounts to a lower bound on $u(\sigma)$ and, therefore, on U.

In one case only is $u(\sigma)$, for such a "random-walk" strategy σ, exactly $e^{\lambda f}$. As is not hard to see and as is no doubt well known, this is when θ is concentrated on (or barely to the right of) a sequence of numbers of the form $\Delta, 0, -\Delta, -2\Delta, \cdots$, and f/Δ is an integer. In this case, f_t is macroscopically greater than 0 with probability 0.

The formula

$$\begin{aligned} E_\sigma(e^{\lambda f_t}|f_t \geq 0) &= E_\sigma(E_\sigma(e^{\lambda f_t}|f_t \geq 0, f_{t-1})|f_t \geq 0) \\ &= E_\sigma(E_\theta(e^{\lambda(z+f_{t-1})}|z + f_{t-1} \geq 0)|f_t \geq 0), \end{aligned} \tag{5}$$

whatever details might be involved in its careful interpretation, suggests that

$$\begin{aligned} \inf_{w \geq 0} \frac{\int_{[w,\infty)} e^{\lambda(z-w)}\, d\theta(z)}{\theta[w, \infty)} &\leq E_\sigma(e^{\lambda f_t}|f_t \geq 0) \\ &\leq \sup_{w \geq 0} \frac{\int_{[w,\infty)} e^{\lambda(z-w)}\, d\theta(z)}{\theta[w, \infty)}. \end{aligned} \tag{6}$$

The inequalities (6) can be established either by careful interpretation of (5) or by means of Theorem 2.12.1.

As de Finetti (1940, pp. 73–74) pointed out, there is one case in which the two extremes of (6) coincide, namely, when the denominators $\theta[w, \infty)$ are of the form $Ae^{-\rho w}$ for some ρ, $\rho > \lambda$. In this case, according to (4) and (6), $u(\sigma) = [1 - (\lambda/\rho)]e^{\lambda f}$, which is not much smaller than $e^{\lambda f}$ if λ/ρ is small.

Obviously, if θ is bounded from above, so that $\theta[B, \infty) = 0$, then (4) with (6) permits the conclusion that $e^{\lambda f} \geq u(\sigma) \geq e^{\lambda f - \lambda B}$. Analysis along the lines of (4) and (6) has also been given for lotteries θ with normal distributions (Wald 1947, Appendix 2.5).

Turn now to the λ-exponential casinoe, the full house under $e^{\lambda f}$. According to Example 7.4.1, for the U of this casinoe,

$$(7) \qquad U(f) = \frac{U_\lambda(f) - U_\lambda(0)}{U_\lambda(1) - U_\lambda(0)} = \frac{e^{\lambda f} - 1}{e^\lambda - 1}$$

for f in $[0, 1]$ and $\lambda > 0$.

Recalling that the two limiting fortunes 0 and 1 ordinarily adopted for a casino are conventional only, rewrite (7) for a casinoe on the interval $[-z, \infty)$ with 0 as the goal, thus.

$$(8) \qquad U(f) = \frac{e^{\lambda f} - e^{-\lambda z}}{1 - e^{-\lambda z}} = e^{\lambda f} \frac{1 - e^{-\lambda(z+f)}}{1 - e^{-\lambda z}}$$

for f in $[-z, 0]$. Note that, once z is so large that $\lambda(z + f)$ is large, the influence of z in (8) is small.

The close isomorphism between the λ-exponential house on the real numbers and the full house under the function f^λ on the nonnegative reals has already been emphasized. For $\lambda \geq 1$, these latter casinoes are, according to Corollary 6.1, the only fair and subfair rich man's casinoes that are full houses. The U of such a house is itself f^λ for $f \leq 1$. Facts for exponential houses have their counterpart for these casinoes; for example, f^λ is an upper bound, often quite good, for the probability of ever reaching 1 starting with an initial fortune f in a subhouse of such a casinoe.

The rest of this section is about applications of the general theory of exponential houses sketched above. Often a given house Γ of interest is a subhouse of a λ-exponential house Γ_λ, in which case $e^{\lambda f}$ is an upper bound for $U(f)$. An evaluation of the largest λ for which $\Gamma \subset \Gamma_\lambda$ is often equivalent to finding the best exponential upper bound for U. The problem typically reduces to the evaluation of $\lambda(\theta)$ for certain lotteries θ. For when $\lambda(\theta)$ exists and is positive, it is the largest λ for which θ belongs to the λ-exponential house. Hence the largest λ for which $\Gamma \subset \Gamma_\lambda$ is the infimum of the numbers $\lambda(\theta)$ as θ ranges over all lotteries available in Γ at any f. When the exact calculation of $\lambda(\theta)$ is difficult, it may be more illuminating to determine a certain simple approximation $I(\theta)$ to $\lambda(\theta)$, as will soon be made apparent.

Suppose first that θ has a normal distribution with mean μ and variance σ^2, $\mu < 0 < \sigma^2$.

$$L_\theta(\lambda) = \frac{1}{\sqrt{2\pi}} \int e^{\lambda f} e^{-(f-\mu)^2/2\sigma^2}\, df; \tag{9}$$

consequently,

$$L_\theta(\lambda) = e^{\lambda(2\mu+\lambda\sigma^2)/2}. \tag{10}$$

Therefore $\lambda(\theta) = -2\mu/\sigma^2$. Call $I(\theta)$ the *inequity* of any θ that has finite positive variance σ^2 and mean μ, whether θ is normally distributed or not, and abbreviate it to $I(\theta)$. As is now evident, when every θ in $\Theta(f)$ (except $\delta(f)$) has a normal distribution with $I(\theta) \geq I > 0$ for some I and all f, then Γ is a subhouse of the I-exponential house. Therefore, $U(f) \leq e^{If}$ for $f \leq 0$. This upper bound is often close, or even exact. For instance, as is known in connection with sequential analysis, if each $\Theta(f)$ contains a fixed θ with normal distribution of negative mean μ and variance σ^2, and if $I(\theta) = I$, then $U(f) = U_I(f)[1 - 0(\mu/\sigma)]$ uniformly in f and I for f in $(-\infty, 0]$ and I in $(0, \infty)$; see (Wald 1947, Appendix 2.5).

Even when θ is not normal but is nearly fair in some sense, $I(\theta)$ promises to be close to $\lambda(\theta)$. One of us, Savage, first heard of this heuristic significance of $I(\theta)$ as an index of the undesirability of a lottery indirectly from Jesse Marcum, and it stimulated Savage's interest in gambling problems. The idea of $I(\theta)$ as an approximation to $\lambda(\theta)$ was published relatively early (de Finetti 1939). De Finetti (1940, p. 64) gives the following mathematical interpretation to this idea:

Let ψ be a lottery of mean 0 and variance 1 such that $L_\psi(\lambda) < \infty$ for some positive λ, and let $\theta = [\sigma\psi + \mu]$ for some positive σ and negative μ. The mean, variance, and inequity of θ are μ, σ, and $I = -2\mu/\sigma^2$.

$$\begin{aligned} L_\theta(\lambda) &= \int e^{\lambda(\sigma f+\mu)}\, d\psi(f) \\ &= e^{\lambda\sigma(\mu/\sigma)} \int e^{\lambda\sigma f}\, d\psi(f) \\ &= e^{xy} L_\psi(x), \end{aligned} \tag{11}$$

where $x = \lambda\sigma$ and $y = \mu/\sigma$. Clearly, $L_\psi(x) > 1$ for $x > 0$. If, for a positive x, $L_\psi(x)$ is finite, then $L_\theta(\lambda) = 1$ for some sufficiently negative y.

$$\text{(12)} \qquad \log L_\theta(\lambda) = xy + \log L_\psi(x) = x(y + x^{-1}\log L_\psi(x)).$$

Aside from the trivial solution $x = 0$, $L_\theta(\lambda) = 1$ if and only if

$$\text{(13)} \qquad y = -x^{-1}\log L_\psi(x).$$

Since $L_\psi(x) = 1 + \frac{1}{2}x^2 + O(x^3)$,

$$\text{(14)} \qquad y = -\tfrac{1}{2}x + O(x^2).$$

According to the implicit function theorem,

$$\text{(15)} \qquad x = -2y + O(y^2).$$

Substituting $\lambda\sigma$ for x and μ/σ for y in (15),

$$\text{(16)} \qquad \lambda(\theta) = \lambda = -\frac{2\mu}{\sigma^2} + \left(\frac{2\mu}{\sigma^2}\right)O\left(\frac{\mu}{\sigma}\right) = \left(1 + O\left(\frac{\mu}{\sigma}\right)\right)I.$$

Therefore, if $-\mu/\sigma$ is small and $f > 0$, then e^{-fI} is at least almost an upper bound to the probability that a sum of independent random variables each distributed according to $[\sigma\psi + \mu]$ ever exceeds f. Incidentally, since the error term of (14) can have either sign, I can be either larger or smaller than $\lambda(\theta)$.

General two-valued lotteries are studied in (de Finetti 1939). The simplest possibilities are the symmetric, two-valued lotteries as in red-and-black, where ϵ has probability w and $-\epsilon$ has probability $\bar{w} = (1 - w)$ for some positive ϵ and for some w in $(0, 1)$. For such a θ, $\mu = w - \bar{w}$, $\sigma^2 = 4w\bar{w}\epsilon^2$, and $I(\theta) = 2\epsilon^{-2}(w^{-1} - \bar{w}^{-1})$. Finally, when $w < 1/2$, $\lambda(\theta)$ is the positive root of

$$\text{(17)} \qquad we^{\lambda\epsilon} + \bar{w}e^{-\lambda\epsilon} = 1;$$

so $e^{\lambda\epsilon} = \bar{w}/w$ and $\lambda = \epsilon^{-1} \log (\bar{w}/w)$. If nothing but contractions of this θ are available, $U(f) \leq e^{\lambda f} = (\bar{w}/w)^{f/\epsilon}$.

As a step in a different problem, Blackwell (1954) extended the preceding conclusion by studying, in effect, a gambling house $\Gamma = \Gamma_{\epsilon,\rho}$ in which the gambler is confined to a sequence of bets for each of which his maximum gain or loss is ϵ and his expected loss is at least a fixed fraction ρ of the most money that has any chance of changing hands. Formally for $[f + \theta] \in \Gamma(f)$ it is necessary and sufficient that $\theta[-\epsilon', \epsilon'] = 1$ for all $\epsilon' > \epsilon$ and that $\mu \leq -\rho\delta$, where μ is the mean of θ and δ is the infimum of the δ' for which $\theta[-\delta', \delta'] = 1$. If θ is concentrated at $-\epsilon$ and ϵ with the highest probability w for ϵ compatible with the initial constraints of the problem, namely, $w = (1 - \rho)/2$, then $\lambda(\theta) = \epsilon^{-1} \log ((1 + \rho)/(1 - \rho)) = \lambda'$ according to (17); therefore θ conserves $e^{\lambda' f}$ and is available in the λ'-exponential house. To see that every θ available in $\Gamma_{\epsilon,\rho}$ is also available in the full house under $e^{\lambda' f}$, evaluate the Laplace transform of θ, thus.

$$(18) \qquad \begin{aligned} L(\lambda) &= \int_{-\delta}^{\delta} e^{\lambda f} \, d\theta(f) \\ &\leq \int_{-\delta}^{\delta} \left\{ \frac{e^{\lambda\delta} + e^{-\lambda\delta}}{2} + \frac{e^{\lambda\delta} - e^{-\lambda\delta}}{2\delta} f \right\} d\theta(f) \\ &\leq \frac{1 - \rho}{2} e^{\lambda\delta} + \frac{1 + \rho}{2} e^{-\lambda\delta}. \end{aligned}$$

Therefore $L(\lambda) \leq 1$ for $\lambda \leq \lambda' \leq (1/\delta) \log [(1 + \rho)/(1 - \rho)]$, and $\Gamma_{\rho,\epsilon}$ is a subhouse of the λ'-exponential house. Consequently, $U(f)$, the optimal probability of reaching the goal 0 from a negative f, is at most

$$(19) \qquad e^{\lambda' f} = \left(\frac{1 + \rho}{1 - \rho} \right)^{f/\epsilon}.$$

This bound is sometimes attained and, when ρ and ϵ are small, can, for all $f < 0$, be nearly attained by repetition of a two-point lottery. The exact nature of the function $U(f)$ is unknown. It is presumably singular and possibly related to the red-and-black casino function. If the gambler is not permitted to let his fortune descend below $-z$, so that he is in a casinoe, then (8) can be applied to obtain an improved bound. The ideas of this paragraph were extended by Weingarten

(1956) to asymmetric bounded gambles in an extension of the main result of (Blackwell 1954).

As was mentioned in Section 1.3, a gambler with an initial fortune of \$1,000 who attempts to attain \$10,000 by bets of \$1 or less at red-and-black is very unlikely to succeed. If he always bets exactly \$1, with probability w of winning, his probability of attaining his goal is exactly

$$(20) \qquad \left(\frac{w}{\bar{w}}\right)^{9,000} \frac{1 - \left(\frac{w}{\bar{w}}\right)^{1,000}}{1 - \left(\frac{w}{\bar{w}}\right)^{10,000}};$$

with infinite credit it is

$$(21) \qquad \left(\frac{w}{\bar{w}}\right)^{9,000}$$

These formulae derivable from (4), (8), and (17), or by the difference-equation approach, have been within the scope of probability theory since almost the first decade of the eighteenth century (De Moivre 1711). For early history of gambler's-ruin, refer to "De Moivre" in the index of (Todhunter 1865). The conclusion of Blackwell in the preceding paragraph shows that the gambler cannot help himself by sometimes using bets even smaller than \$1.

For a numerical example, substitute $w = 9/19$—which corresponds to the version of roulette in Nevada—into (21), and calculate a very crude upper bound to the probability of success, thus.

$$(22) \qquad \left(\frac{w}{\bar{w}}\right)^{9,000} = \left(\frac{9}{10}\right)^{9,000} = \left(1 - \frac{1}{10}\right)^{9,000} < e^{-900} < 10^{-300}.$$

With bold play, the gambler can attain his goal with probability $Q_w(1/10)$, in the notation of Chapter 5. As pointed out in Section 5.2, $Q_w(f)$ is a rational function of w for each rational f, obtainable by repeated application of the functional equations (4.2.1). In particular,

$$(23) \qquad Q_w(1/10) = w^4(1 + \bar{w})(1 - w^2\bar{w}^2)^{-1}.$$

For $w = 9/19$, $.081 < Q_w(1/10) < .082$, an improvement over 10^{-300}.

9
THREE PARTICULAR KINDS OF CASINOS

1. INTRODUCTION.

Each of the three main sections of this chapter explores a certain measure of subfairness.

2. INCOME-TAX CASINOS

As a preliminary to his construction of an example of a fair lottery that generates a subfair casino, Donald Ornstein found that a useful measure of the subfairness of a nontrivial lottery with a finite mean is the ratio of the expected value of its positive part to that of its negative part. Such a lottery is fair if and only if this ratio is 1. If a gambler purchases a fair lottery but is required to return a fraction t of his winnings to the house—a fraction that can be viewed as a tax on winnings or "income"—he has, in effect, purchased a lottery for which the ratio is $1 - t$. For this reason,

$$t(\theta) = 1 - \frac{\int_0^\infty w \, d\theta(w)}{\int_{-\infty}^0 - w \, d\theta(w)} \tag{1}$$

will be called the *tax* of any lottery θ for which the fraction is well defined.

Example 1. If θ is a two-valued lottery that gains $\bar{r}$ and loses r with probabilities w and $\bar{w}$, $t(\theta) = 1 - (w\bar{r}/\bar{w}r)$.

Of course, $t(\theta)$ does not exceed 1, and (1) is conveniently rewritten as

$$(2) \qquad (1 - t(\theta)) \int_{-\infty}^{0} - w \, d\theta(w) = \int_{0}^{\infty} w \, d\theta(w).$$

Clearly, $t(\theta)$ is positive, zero, or negative according as θ is subfair, fair, or superfair. Only the subfair case is of interest at present. Accordingly, for t in $(0, 1]$ the *income-tax casino* Γ_t is the casino in which $\Theta_t(f)$ for $f > 0$ consists of all lotteries θ on $[-f, \infty)$ for which

$$(3) \qquad (1 - t) \int_{-\infty}^{0} - w \, d\theta(w) \geq \int_{0}^{\infty} w \, d\theta(w).$$

THEOREM 1. For $0 < t \leq 1$, the income-tax casino Γ_t is an inclusive casino.

The following theorem and preliminary lemma, which were observed by Ornstein, are recorded here with his permission. Let u be the usual two-valued utility and $U_t(f)$ the probability of reaching 1 from f under optimal play in the income-tax casino Γ_t.

LEMMA 1

(*a*) Γ_t is leavable and closed under composition.

(*b*) $U_t(f)$ is the supremum of γu over all $\gamma \in \Gamma_t(f)$.

(*c*) For $\gamma \in \Gamma_t(f)$, $\gamma u = U_t(f)$ if and only if, for each $\epsilon > 0$, $\gamma[1, 1 + \epsilon) = U_t(f)$, $\gamma[0, \epsilon) = \bar{U}_t(f)$, and $\bar{f}\gamma[1, 1 + \epsilon) = (1 - t)f\gamma[0, \epsilon)$.

(*d*) $\bar{f}U_t(f) = (1 - t)f\bar{U}_t(f)$.

THEOREM 2. $U_t(f) = (f - tf)/(1 - tf)$.

In any casino, $U(f)/\bar{U}(f)$ might be termed the *effective odds* for that casino at f. If $U(f)/\bar{U}(f) = f/\bar{f}$, the odds are *fair;* only fair casinos have fair odds. In the t-income-tax casino,

$$\frac{U_t(f)}{\bar{U}_t(f)} = (1 - t)\left(\frac{f}{\bar{f}}\right);$$

that is, the odds are a fixed fraction $(1 - t)$ of the fair odds.

COROLLARY 1. γ conserves U_t in Γ_t at an f in $[0, 1]$ if and only if

$$\lim_{\epsilon \to 0} \gamma[0, \epsilon) + \gamma(f - \epsilon, f + \epsilon) + \gamma(1 - \epsilon, 1 + \epsilon) = 1,$$

and

$$\bar{f} \lim_{\epsilon \to 0} \gamma(1 - \epsilon, 1 + \epsilon) = (1 - t)f \lim_{\epsilon \to 0} \gamma[0, 0 + \epsilon).$$

Income-tax casinos have several applications to primitive casinos.

THEOREM 3. The primitive-casino $\Gamma_{w,r}$ is a subhouse of the income-tax casino Γ_t with $t = 1 - (w\bar{r}/\bar{w}r)$, for $0 < w < r < 1$.

In view of Theorem 3, the utility $U_{w,r}$ of $\Gamma_{w,r}$ is majorized by U_t. In particular, $U_{w,r}(r) \leq U_t(r) = w$. Since the two-point gamble γ, $\gamma\{1\} = w$, $\gamma\{0\} = \bar{w}$, is in $\Gamma_{w,r}(r)$, $U_{w,r}(r) \geq w$. The conclusion $U_{w,r}(r) = w$ was derived in Section 6.3 by combinatorial arguments. We have not found a noncombinatorial argument showing even that $U_{w,r}(r^2) = w^2$.

That bold strategies are optimal for $\Gamma_{w,r}$—equivalently, that $S_{w,r}$ is $U_{w,r}$—is not used in the proofs of the following results about $S_{w,r}$.

THEOREM 4. $U_t(f) \geq S_{w,r}(f)$ for f in $[0, 1]$, if $t = 1 - (w\bar{r}/\bar{w}r)$ and if $0 < w \leq r < 1$. Equality holds at and only at $f = 0$, r, and 1.

Proof: Let $\sigma(f)$ be the bold strategy at f. Since $S_{w,r}(f) = \sigma(f)u$, and since $\theta(f)$ is available in Γ_t at f,

$$U_t(f) \geq \sigma(f)U_t \geq \sigma(f)u = S_{w,r}(f). \tag{4}$$

In view of Corollary 1, the initial gamble of $\sigma(f)$ does not conserve U_t at f for any f in $[0, 1]$ except $f = 0$, r, and 1. Therefore, according to Corollary 3.3.4, the first inequality in (4) is strict except at 0, r, and 1. ◆

If the lotteries of two primitive casinos have the same tax, then neither primitive casino is more favorable than the other for every initial f.

COROLLARY 2. If $w\bar{r}/\bar{w}r = w'\bar{r}'/\bar{w}'r'$, then neither $S_{w,r}$ nor $S_{w',r'}$ majorizes the other unless both are fair (that is, $w = r$ and $w' = r'$) or $w = w'$ and $r = r'$.

A phenomenon mentioned in Section 6.7 is exhibited immediately following this lemma, which is implied by Theorem 7.3.2.

LEMMA 2. If, on an interval I, U_1 and U_2 are twice differentiable, and $\dot{U}_1$ and $\dot{U}_2$ are both positive or both negative, then Γ_1, the full house under U_1 on I, includes Γ_2, the full house under U_2 on I—equivalently, U_1 is excessive for Γ_2—if and only if $\ddot{U}_2/\dot{U}_2$ majorizes $\ddot{U}_1/\dot{U}_1$.

THEOREM 5. There are pairs of casinos Γ_1 and Γ_2 such that $U_1 \geq U_2$ and the U of the union of Γ_1 and Γ_2 (which must be at least U_1) is not U_1.

Proof: Let Γ_1 and Γ_2 be the income-tax casino for $t = 1/2$ and the full-house casino under f^2. Then $U_1(f) - U_2(f) = f(2 - f)^{-1} - f^2 = f\bar{f}^2(2 - f)^{-1}$; so U_1 does majorize U_2. Since $\ddot{U}_1(f)/\dot{U}_1(f) = 2(2 - f)^{-1}$ and $\ddot{U}_2(f)/\dot{U}_2(f) = f^{-1}$, $\ddot{U}_2/\dot{U}_2$ does not majorize $\ddot{U}_1/\dot{U}_1$. Lemma 2 now implies that U_1 is not excessive for the full house under f^2, and it is therefore not the U of any Γ which includes that full house (Theorem 2.14.1). ♦

In view of Theorem 2, the linear fractional transformations U_t satisfy the casino inequality (4.2.1) (as was asserted after (4.6.7) in slightly different notation).

The full houses under the casino functions U_t are not casinos, as is easily seen with the help of Theorem 8.5.1.

The income-tax casinos are permitted houses. This suggests an approach different from Lemma 1 to the calculation of U_t and illustrates the use of Theorem 7.5.2. Consider $D = D_t$ the *dog-leg function* of *slope ratio* $1 - t$, where

$$D(g) = \begin{cases} g & \text{for } g \geq 0, \\ (1 - t)g & \text{for } g \leq 0. \end{cases} \tag{5}$$

The income-tax casino Γ_t is easily seen to be the house permitted by the family

$$v_f(g) = D(g - f). \tag{6}$$

For each f in $[0, 1]$ let

$$w_f = a(f)v_f + b(f), \tag{7}$$

where $a(f)$ and $b(f)$ are such that $w_f(0) = 0$ and $w_f(1) = 1$. According to elementary computation,

$$a(f) = \frac{1}{1 - tf}, \tag{8}$$

$$b(f) = \frac{f - tf}{1 - tf}. \tag{9}$$

Since b is convex in f for $f \leq 1$, and the graph of each w_f is a polygon inscribed in that of b, $b \leq w_f$ for all f. Therefore, according to Theorem 7.5.2, b is excessive for the income-tax casino Γ_t. Of course b is at least u, and Theorem 2.12.1 implies that b is at least U_t. Since, for each f in $[0, 1]$, there is a two-point $\gamma \in \Gamma_t(f)$ carried by the points 0 and 1 for which $\gamma\{1\} = \gamma w_f = b(f)$, $b(f) \leq U_t(f)$. All in all, $b(f) = U_t(f)$ for $0 \leq f \leq 1$, which has already been reported in Theorem 1.

More generally, Theorem 7.5.6 (or the more detailed Theorem 7.3.4 on which it is based) provides an avenue for solving the gambling problem defined by Γ_t and a general u.

The gambler cannot diminish the tax by changing the scale of a lottery or by buying several lotteries simultaneously:

LEMMA 3. For any positive number a and subfair random variable z, the tax of az equals that of z. The tax of the sum of subfair random variables is at least the minimum of their taxes.

Proof: The first assertion is obvious from the definition of tax. The second follows from the easy inequality

$$D_t(f + g) \leq D_t(f) + D_t(g). \quad \blacklozenge \tag{10}$$

THEOREM 6. If $a_1, \cdots, a_n$ are positive numbers and $z_1, \cdots, z_n$ are subfair random variables, then the tax of $a_1z_1 + \cdots + a_nz_n$ is at least the minimum of the taxes of the z_i.

The analogue of even the first part of Lemma 3 is true neither for the inequity nor for the critical λ of Section 8.7, but the second part of Lemma 3 does apply to these measures of subfairness if z_1 and z_2 are independent.

3. THE CASINO THAT TAKES A CUT OF THE STAKE

Whenever people talk about a lottery or other game of "pure chance", like a specific bet at roulette, they are apt to take for granted that *the* measure of subfairness of the lottery to the customer and of its profitability to the house is what we shall here call the cut. The *stake* of a lottery θ is the (essential) supremum of the losses under θ, that is, the infimum of the nonnegative numbers s for which $\theta[-s, \infty) = 1$. If θ has positive stake s and is subfair, the *cut* c is the ratio of the negative of the mean m of θ to the stake. In these terms, the purchase of a lottery ticket (or insurance policy) of price (or premium) s can be envisaged thus. A fraction c of s, the cut (or loading), is paid by the gambler as a fee to the house (or insurance company); he is paid back according to the fair lottery $\theta - m = \theta + sc$.

The cut of a lottery (or the loading of a policy) may be a parameter of economic importance in some contexts. Two easy facts suggestive of such importance are that the analogue of Theorem 2.6 applies to the cut as well as to the tax and that the tax of a lottery cannot be less than the cut. These facts, especially the former, may be largely responsible for the widespread feeling that the cut cannot be reduced by means of some policy; see, for example, (Scarne 1961, p. 385). Nevertheless, when the gambler can buy lotteries consecutively, each in the light of the outcome of the last, he may be able to construct lotteries of arbitrarily low cut from lotteries of cut arbitrarily close to 1.

Example 1. In a subfair primitive casino characterized by w and r, the cut of all available lotteries (aside from $\delta(0)$) is $c = (\bar{w}r - w\bar{r})/r = (r - w)/r = 1 - (w/r)$. A gambler whose initial fortune is $1 - \bar{r}^n$ and who plays boldly constructs for himself, in n moves, a simple lottery with $r' = 1 - \bar{r}^n$, $w' = 1 - \bar{w}^n$ (Section 6.2). The cut of this constructed lottery is $(r' - w')/r' = (\bar{w}^n - \bar{r}^n)/(1 - \bar{r}^n)$, which approaches 0 in n. Similarly, the gambler in the primitive casino who starts at r^n must construct for himself a lottery with the large cut $1 - (w/r)^n$ if he is to play optimally.

Example 2. In any subfair casino with continuous U, the gambler who uses a nearly optimal policy beginning at f in $(0, 1)$ constructs for himself a lottery of cut scarcely larger than $1 - (U(f)/f)$, which is arbitrarily small for f near 1.

Example 3. The theory of uniform roulette (Section 6.9) led to the interesting example of two casinos in which all gambles have the same cut $1 - (w/r)$ but with the casino function of one strictly larger than that of the other throughout $(0, 1)$, which contrasts with Corollary 2.2.

Though the minimum cut of lotteries directly offered by a gambling house does not ordinarily apply to lotteries that the gambler can construct for himself within the gambling house, it does express some constraint on what the gambler (with limited funds) can do. Any subfair gambling house on $[0, \infty)$, the lotteries of which have cut at least c, is a subhouse of the casino in which each $\Theta(f)$ consists of all lotteries θ on $[-f, \infty)$ that are not superfair and for which $c(\theta) \geq c$ if $s(\theta) > 0$. Therefore the U of this casino, the *casino that takes a cut of* c, gives an upper bound of the probability of attaining 1 from an initial fortune in any such gambling house. The main object of this section is to calculate U.

Of course, if a simple lottery of cut c and stake s leads from $f < 1$ to 1 with probability w, then

$$(1) \qquad w = \frac{(1 - c)s}{\bar{f} + s}; \qquad \bar{w} = \frac{\bar{f} + cs}{\bar{f} + s}.$$

If a gambler at f plays boldly, staking f on such a lottery, $w = f - cf$, which is obviously poor for f near 1. It is easy to find better strategies. Trial and error suggest that a very good one is to stake $(1/n)$th of the initial fortune f (if $f < 1$) on a simple lottery of cut c that leads to 1 with probability w_1 and to $(n - 1)f/n$ with probability $\bar{w}_1$ for some large integer n. Then if this gamble does not attain the goal, f/n is staked again with cut c and with probability w_2 of attaining the goal and probability $\bar{w}_2$ of falling to $(n - 2)f/n$, continuing until the goal is attained or the fortune 0 is reached. The successive fortunes, if the goal is not attained, are $f_i = (n - i)f/n$ $(1 \leq i \leq n)$, and

$$w_i = \frac{\bar{c}f}{n\bar{f} + if}. \tag{2}$$

The logarithm of the probability that the goal is not attained with this strategy is

$$\begin{aligned} \sum_1^n \log(1 - w_i) &= -\sum_1^n w_i + \sum_1^n o(w_i) \\ &= -\bar{c}\sum_1^n \frac{f/n}{\bar{f} + (i/n)f} + o(1/n)n \\ &= -\bar{c}\int_0^f \frac{dz}{\bar{f} + z} + o(1) \\ &= \bar{c}\log\bar{f} + o(1). \end{aligned} \tag{3}$$

Therefore the probability of attaining the goal with this strategy is, for large n, about

$$1 - (1 - f)^{(1-c)}. \tag{4}$$

Many other strategies, and even stationary families of strategies, are about as good; for example, the f_i can be replaced by any descending sequence, with very small steps, that reaches, or converges to, 0. Since (4) can nearly be attained, U majorizes the function W given by (4) for $f \leq 1$ and equal to 1 for $f \geq 1$. It will next be shown by means of the basic Theorem 2.12.1 that W majorizes U and therefore is U.

To see that $W \geq U$, the only nontrivial step is to verify that $\gamma W \leq W(f)$ for $\gamma \in \Gamma(f)$ and $f \in (0, 1)$. If s is the stake of γ, then γ is (all but) carried on the interval $[f - s, \infty)$, and the function W is majorized there by the linear function L for which $L(f - s) = W(f - s)$ and $L(1) = W(1) = 1$. Therefore, $\gamma W \leq \gamma L \leq L(f)$ minus cs times the slope of L, or $L(f - s)$ plus $(s - cs)$ times the slope of L.

Analytically,

$$\begin{aligned}(5) \qquad \gamma W &\leq W(f - s) + s(1 - c)\,\frac{1 - W(f - s)}{1 - (f - s)} \\ &= W(f - s) + s\dot{W}(f - s) \\ &< W(f),\end{aligned}$$

where the final inequality follows from the strict convexity of (4) for c in $(0, 1)$. Therefore $W \geq U$.

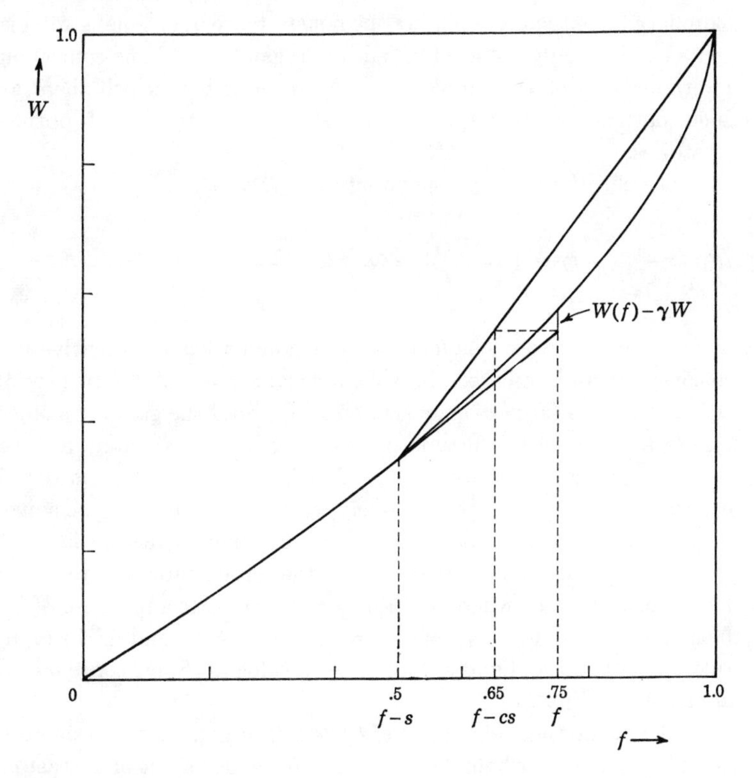

Figure 1 Illustration of (5) with $f = .75$, $c = .4$, $s = .25$.

THEOREM 1. There is at each f in Γ_c a stationary optimal strategy, and yet no stationary family of optimal strategies is available.

Proof: For f in $(0, 1)$, let σ_0 be diffuse on $f - n^{-1}$. Let $\sigma_1[f']$ be stationary and n^{-1}-optimal for $f' = f - n^{-1}$. The σ thus determined is optimal at f.

Nonetheless, if $\gamma(f)$ determined a stationary family of optimal strategies, then $U(f) = \gamma(f)U$ (according to (3.3.1)); so the inequality (5) implies that the stake of $\gamma(f)$ is necessarily 0, that is, $\gamma(f)$ is trivial. Since u is excessive for $\gamma(f)$ and $u \geq u$, Theorem 2.12.1 implies that $u \geq U$, which is a contradiction. ◆

According to (5), no γ in $\Gamma_c(f)$ other than $\delta(f)$, which is countably additive on the Borel sets or even on the intervals, conserves U_c at f. Corollary 3.3.4 therefore implies the nonexistence of optimal strategies that employ only countably additive gambles. Time-continuous reformulations of the problem can be suggested in which there are such optimal strategies and even stationary families of optimal strategies.

The effective cut corresponding to $U(f)$ is

$$c(f) = 1 - \frac{U(f)}{f} = \frac{\bar{f}^{\bar{c}} - \bar{f}}{f} < \frac{1 - \bar{c}f - \bar{f}}{f} = c, \tag{6}$$

which reflects the obvious fact that the gambler has no incentive in Γ to construct for himself gambles of cut as large as c. According to (6), $c(f)$ is nearly as large as c for very small f. So if the gambler's objective is to increase his fortune manyfold, c is approximately a direct measure of the subfairness of Γ. The fraction ϕ of the cut that is effective under optimal play, namely, $c(f)/c$, is defined on the square $0 < f < 1$, $0 < c < 1$ and is decreasing in f and increasing in c. It extends to a continuous function on the closed unit square except for the corner $(1, 1)$, where its value is indeterminate and where $\phi(f, c)$ behaves like $\bar{f}^{\bar{c}}$. Clearly, $\phi(f, 1) = 1$, $\phi(1, c) = 0$, and L'Hospital's rule gives $\phi(0, c) = 1$ and $\phi(f, 0) = -(\bar{f}/f) \log \bar{f}$. Some contours of ϕ are shown in Fig. 2.

The casino that takes a cut of c has a natural dual, the casino generated by all subfair bounded lotteries θ for which a cut of at least $c/\bar{c}$ is taken from the (essential) supremum of the gains; that is,

$$-m(\theta) \geq \left(\frac{c}{\bar{c}}\right) \sup \theta. \tag{7}$$

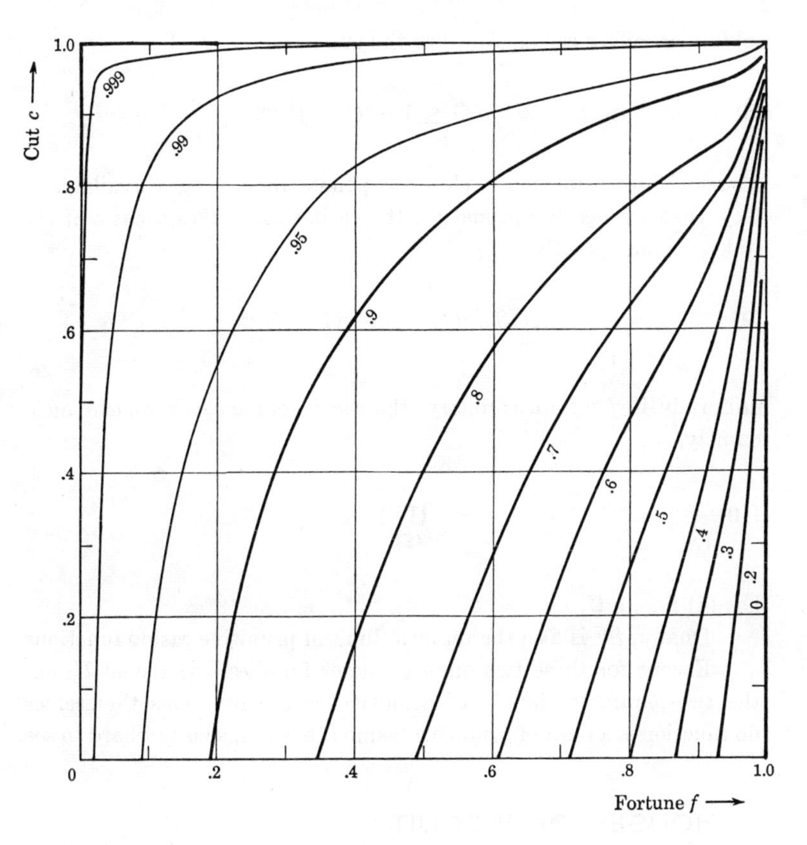

Figure 2 Contours of ϕ.

The demonstrations of various facts about this dual of the casino that takes a cut of the stake are too similar to corresponding ones about the original casino to be repeated explicitly. Here a typical attractive gamble is to stake the whole of f, for f in $(0, 1)$, on a high probability of inching forward a bit. The U of this casino is the dual of the U of the casino that takes a cut c of the stake, namely, $f^{1/\bar{c}}$. These casino functions were encountered near the end of Section 8.6 as the casino functions of the only subfair, rich man's casinos that are full houses.

As Lawrence Jackson helped us see, this section has implications for primitive casino functions. Since every nontrivial lottery in the subfair primitive casino $\Gamma_{tw,tr}$ has the cut $c = 1 - (w/r)$,

$$S_{tw,tr}(f) \le 1 - (1 - f)^{1-c}. \tag{8}$$

Moreover, for t sufficiently close to 0, there are strategies available in $\Gamma_{tw,tr}$ that are nearly optimal for the casino that takes a cut c of the stake. Consequently,

$$\lim_{t \to 0} S_{tw,tr}(f) = 1 - (1 - f)^{1-c}, \tag{9}$$

uniformly in f. As a corollary, the *continuous uniform-roulette table*, namely,

$$\bigcup_{0<t\le 1} \Gamma_{tw,tr}, \tag{10}$$

has (4) for its U.

Dually, $f^{1/\bar{c}}$ is also the uniform limit of primitive casino functions.

Except for these two one-parameter families, the trivial U, and the two-parameter family of primitive casino functions themselves, no function is a limit of primitive casino functions, as is not hard to see.

4. HOUSES OF INEQUITY

In Section 8.7 the inequity of a lottery θ, that is, $I(\theta) = -2\mu/\sigma^2$, was seen to be an index of the undesirability of θ in that $I(\theta)$ often approximates the exponential measure of subfairness $\lambda(\theta)$. This section shows exactly how great a limitation a given inequity imposes.

THEOREM 1. Suppose that for some positive constant α every available lottery θ is such that

$$-\int z\, d(\theta) \ge \alpha \int z^2\, d\theta(z) \tag{1}$$

or even only such that the ratio of the negative of the mean to the variance (that is, $\frac{1}{2}I(\theta)$) is at least α, then under any policy for which the probability of losing more than l is 0, the probability of gaining g is at most

$$\frac{l}{g + l} \cdot \frac{1}{1 + \alpha g}. \tag{2}$$

This bound can nearly be attained by a succession of two-valued lotteries each of which has a very small stake and a positive probability of reaching the goal.

Proof: It is not difficult to see, say by an argument like one given in Section 3, that the goal is nearly attained in the manner specified with lotteries that satisfy (1) with equality. Since the condition $-\mu/\sigma^2 \geq \alpha$ is plainly less severe than condition (1), all that remains s to establish that (2) is a bound under the condition $-\mu/\sigma^2 \geq \alpha$.

It suffices to show, for the casinoe where

$$\Theta(f) = \{\theta\colon I(\theta) \geq 2\alpha, \theta[-f, \infty) = 1\}$$

with u the indicator of $[t, \infty)$ for arbitrary positive t, that U is majorized by Q, where

$$Q(f) = \begin{cases} \dfrac{f}{t[1 + \alpha(t - f)]} & \text{for } 0 \leq f \leq t, \\ 1 & \text{for } t \leq f < \infty. \end{cases} \tag{3}$$

For $f < t$ let R_f be the quadratic function whose graph is tangent to the graph of Q at the point $(f, Q(f))$ and for which $R_f(t) = 1$. As is easily verified,

$$R_f(f + x) = Q(f) + \dot{Q}(f)(x + \alpha x^2), \tag{4}$$

and as is less easy to verify,

$$Q(f + x) \leq R_f(f + x) \qquad \text{for } f + x \geq 0. \tag{5}$$

If γ is a gamble available at $f + \Delta$ with mean f and variance σ^2, then

$$\text{(6)} \qquad \begin{aligned} \gamma Q \le \gamma R_f &= Q(f) + \alpha\sigma^2 \dot{Q}(f) \\ &\le Q(f) + \Delta \dot{Q}(f) \\ &\le Q(f + \Delta). \quad \blacklozenge \end{aligned}$$

The especially simple and interesting limiting case of Theorem 1 with $l = -\infty$ was emphasized in (Dubins and Savage 1965) and applied in (Dubins and Freedman 1965).

What bound replaces (2) if "variance" is replaced by "standard deviation" in the hypothesis of Theorem 1?

10 ONE-LOTTERY STRATEGIES

1. INTRODUCTION. This chapter is basically an application of part of the familiar theory of sums of independent and identically distributed random variables.

A casino Γ has a superfair lottery θ available at 1 if and only if Γ is superfair; equivalently, if and only if, for the usual u, U is the indicator of $(0, \infty)$. The proof of this in Section 4.4 is indirect; so there is some interest in finding a stationary family of optimal strategies based on θ. One natural possibility, the *θ-proportional family,* is always to use $[f\theta + f]$ for $f < 1$ and $\delta(f)$ for $f \geq 1$, that is, to use θ in proportion to your wealth until the goal is achieved. Somewhat surprisingly, this is not always optimal, even for quite ordinary lotteries θ. What is needed for it to be optimal is roughly that $\int \log (1 + z)\, d\theta(z) \geq 0$; when this integral is undefined, the θ-proportional family may or may not be optimal. Though the integral can be negative for θ itself, it is bound to be positive for the contraction $[s\theta]$ for sufficiently small positive s; so the $[s\theta]$-proportional family is optimal for small s. This result, the main one of this chapter, will be used as two technical points in the next chapter.

Those who do not wish to enter into the formal proof and the refinements of the preceding paragraph can skip or skim this chapter.

To replace multiplication by addition and to take advantage of the established theory of sums of independent and identically distributed random variables, study of proportional families will be founded upon that of translational families, defined as follows: If ϕ is a lottery on $[-\infty, \infty)$, the *ϕ-translational family* of strategies $\bar{\sigma}$ is the family of strategies generated by the gambles $\bar{\phi}(f) = [\phi + f]$ if $f < 0$ and $\delta(f)$ if $f \geq 0$. It is to be understood that $\bar{\phi}(-\infty) = [\phi - \infty] = \delta(-\infty)$. The utility u of interest now and until Section 5 is the indicator of $[0, \infty)$, and the main question, which we cannot—and need not—answer completely, is "For which ϕ is $u(\bar{\sigma}(f)) = 1$ for all $f > -\infty$?"

For a useful reformulation, consider the gambling house Γ_ϕ for which $\Gamma_\phi(f)$ consists only of $[\phi + f]$. This house is leavable only in the trivial case that $\phi = \delta(0)$. It has only one strategy available at each f; this stationary family σ^* is the family generated by $\phi^*(f) = [\phi + f]$ at all f in $[-\infty, \infty)$.

THEOREM 1. $\bar{\sigma}(f)$ is an optimal strategy at f of Γ_ϕ^L, the leavable closure of Γ_ϕ; $u(\bar{\sigma})$ is the U of Γ_ϕ^L and of Γ_ϕ.

Proof: The two assertions are plainly equivalent to each other. The first is an instance of Theorem 3.9.5, the proof of which we did not give. But the second is easily proved directly, thus: Plainly, $u(\bar{\sigma})$ majorizes u. According to Theorem 3.2.1, $\phi(f)u(\bar{\sigma}) = u(\bar{\sigma}(f))$; the basic Theorem 2.12.1 now applies. ◆

2. LIMITED PLAYING TIME

Let U_m $(m = 0, 1, \cdots, \omega)$ be defined for Γ_ϕ^L as in Section 2.15. It will ultimately be shown that U of Γ_ϕ^L and Γ_ϕ is U_ω. All these functions are plainly 1 for $f \geq 0$.

LEMMA 1. $u(\bar{\sigma}(f), t)$ is nondecreasing in the stop rule t.

Proof: Apply Theorem 3.4.7. ◆

LEMMA 2. $U_m(f) = u(\bar{\sigma}(f), m)$.

LEMMA 3. For $0 < m < \omega$,

$$U_m(f) = \sigma^*(f)\{h\colon \max(f, f_1, \cdots, f_m) \geq 1\}.$$

The next two lemmas are instances of Theorem 8.2.4 and of

(8.3.1), (8.3.2), and (8.3.3), except for the minor detail that $-\infty$ was not included as a fortune there.

LEMMA 4. $U_m(f)$, $U_\omega(f)$, and $U(f)$ are nondecreasing in m and in f for $0 \leq m < \omega$ and $-\infty \leq f < \infty$.

LEMMA 5. $U_{m+n}(f+g) \geq U_m(f)U_n(g)$, $U_\omega(f+g) \geq U_\omega(f)U_\omega(g)$, and $U(f+g) \geq U(f)U(g)$ for $0 \leq m < \omega$ and $-\infty \leq f, g < \infty$.

LEMMA 6. $U_\omega(nf) \geq U_\omega^n(f)$, and $U(nf) \geq U^n(f)$.

THEOREM 1. $U_\omega(f) = 1$ either for all f in $(-\infty, 0)$ or for none; similarly for $U(f)$.

THEOREM 2. $U_0(f)$ is the indicator of $[0, \infty)$, $U_1(f) = \phi[-f, \infty]$ for $f < 0$, and more generally,

$$(1) \qquad U_{m+1}(f) = \begin{cases} \int U_m(g+f)\, d\phi(g) & \text{for } f < 0, \\ 1 & \text{for } f \geq 0. \end{cases}$$

3. CONVENTIONAL LOTTERIES

If the lottery ϕ that generates σ^* is *conventional*, that is, countably additive on the Borel sets of $[-\infty, \infty)$, then the distribution of $(f_1, \cdots, f_m)$ under σ^* on the Borel sets of the m-fold Cartesian product of $[-\infty, \infty)$ is the same as that of $(f + s_1, \cdots, f + s_m)$, where

$$(1) \qquad s_m = \sum_1^m y_j,$$

and the y_j are random variables independently distributed according to ϕ.

The inclusion of $-\infty$ as a possible value of y_j (to provide a logarithm for 0) is a little unusual but not unheard of. The reader will easily see that it does not interfere.

When ϕ is conventional, $U_m(f)$ is the probability that at least one of the partial sums $0, s_1, \cdots, s_m$ is as large as $-f$, according to

Lemma 2.3. And, in conventional terms, $U_\omega(f)$ is the probability that 0 or some s_n is as large as $-f$. Thus, much of the literature on sums like the s_n can be brought to bear. In the last decade there have been studies of $U(f)$ and $U_m(f)$ for conventional ϕ that go far beyond what is needed in this book. A central work in this literature is (Spitzer 1956). Theorem 4.1 of that paper includes the following lemma. You may find the alternative demonstration presented below interesting or convenient.

LEMMA 1. lim sup s_n is a constant almost surely. This constant can be only ∞, 0, or $-\infty$. It is 0 if and only if $y_j = 0$ almost surely; it is ∞ $(-\infty)$ if and only if sup $s_n > 0$ (inf $s_n < 0$) almost surely.

Proof: Apply a variant of the zero-one law (Hewitt and Savage 1955, Theorem 11.3) to see that lim sup s_n is some constant c almost surely. Suppose $0 < c < \infty$. Some s_n is almost sure to exceed $c/2$; let n_1 be the first n for which this happens. Then some $s_n' = s_{n_1+n} - s_{n_1}$ is almost sure to exceed $c/2$, since the s_n' have the same distribution as the s_n (Doob 1953, p. 145). Proceeding in this way, construct $s_{n_1}, s_{n_1+n_2}, \cdots$ converging to ∞. Therefore lim sup $s_n = \infty$, contradicting the assumption $0 < c < \infty$. Likewise, $-\infty < c < 0$ is impossible. As is not difficult to show, if $c = 0$, then $y_j = 0$ almost surely. ♦

LEMMA 2. For a conventional ϕ, $U_\omega(f) = 1$ for $-\infty < f < 0$ if and only if lim sup $s_n = \infty$ almost surely, or, equivalently, if and only if sup $s_n > 0$ almost surely. Otherwise, $U_\omega(f) < 1$ for $-\infty < f < 0$, and $s_n = 0$ for all n almost surely or s_n converges to $-\infty$ almost surely.

Proof: Apply Lemma 1 and the above remark that $U_\omega(f)$ is the conventional probability that 0 or some s_n is at least $-f$. ♦

Theorems 1, 2, and 3 will be proved in this section for conventional lotteries only. But the proofs will be extended to all lotteries in the next section.

THEOREM 1. If $U_\omega(f) < 1$ for $-\infty < f < 0$, then $U_n(f)$ converges uniformly to $U_\omega(f)$.

Proof: For $f \geq 0$, $U_n(f) = U_\omega(f) = 1$. For $f < 0$, $U_\omega(f) -$

$U_n(f)$ is the probability that, for some $m > n$, but for no $m \leq n$, $s_m \geq -f$. This is at most the probability that, for some $m > n$, $s_m > 0$, which approaches zero in n. ◆

THEOREM 2. $U = U_\omega$.

Proof: The conclusion is obvious if $U_\omega(f) = 1$ for $-\infty < f < \infty$. In the other case, it follows immediately from the preceding theorem and part (*b*) of Theorem 2.15.5. ◆

Necessary and sufficient conditions on ϕ, the distribution of y_j, for lim sup s_n to be infinite, that are convenient for this chapter, are not known to us. (See, however, (Spitzer 1956, Theorem 4.1).) If Ey_j is well defined, perhaps infinite, then Kolmogorov's law of large numbers (Loève 1955, page 423; Doob 1953, page 141) supplemented by an interesting theorem of Chung and Fuchs (1951, Theorem 5) readily supplies a convenient condition.

LEMMA 3. If $Ey_j > 0$, s_n converges to ∞ almost surely; if $Ey_j < 0$, s_n converges to $-\infty$ almost surely; if $Ey_j = 0$, and y_j is not almost identically 0, then lim sup $s_n = \infty$ and lim inf $s_n = -\infty$ almost surely. If Ey_j^+ and Ey_j^- are both infinite, then lim sup s_n can be almost surely ∞ and can be almost surely $-\infty$.

Proof: The discussion above suggests the proof for all except the last sentence. If y_j has a standard Cauchy distribution, then lim sup $s_n = \infty$. But if, for example, prob $(y_j = -\infty) > 0$, then lim sup $s_n = -\infty$. ◆

There are also examples with prob $(y_j = -\infty) = 0$, Ey_j^+ and Ey_j^- both infinite, and lim sup $s_n = -\infty$ (Derman and Robbins 1955).

THEOREM 3. For each lottery ϕ on $[-\infty, \infty)$:

(*a*) If $\phi(\epsilon, \infty) = 0$ for all positive ϵ, then $U(f) = 0$ for $-\infty < f < 0$.

(*b*) If $\phi(\epsilon, \infty) > 0$ for some positive ϵ, and $\int f\, d\phi(f) \geq 0$, then $U(f) = 1$ for $-\infty < f < 0$.

(*c*) If $\int f\, d\phi(f) < 0$, then $U(f) < 1$ for $-\infty < f < 0$.

(*d*) If $\int f\, d\phi(f)$ is not defined, then either $0 < U(f) < 1$ for $-\infty < f < 0$ or $U(f) = 1$ for $-\infty < f < 0$. Both cases are possible, and the first obtains if $\phi\{-\infty\} > 0$.

Theorem 3 is the main goal of this section. It has been so expressed as to make the hypothesis that ϕ is conventional superfluous, as will be shown in the next section.

4. GENERAL LOTTERIES

This section extends to arbitrary lotteries on $[-\infty, \infty)$ the main conclusions about conventional lotteries in the preceding section.

One possible peculiarity of a nonconventional ϕ on $[-\infty, \infty)$ is that it can be *partially remote*, that is, $\phi(c, \infty)$ can be uniformly positive for finite c. The facts catalogued in the next theorem are easy to prove from first principles.

THEOREM 1. If ϕ is partially remote:

(*a*) $\int f\, d\phi(f)$ is infinite or undefined.

(*b*) If $\phi\{-\infty\} = 0$, $U(f) = 1$ for $-\infty < f < 0$.

(*c*) If $\phi\{-\infty\} > 0$, $\int f\, d\phi(f)$ is undefined, $0 < U(f) < 1$ for $-\infty < f < 0$, and U_n converges uniformly to U.

In harmony with Section 4.5, two lotteries ϕ and ϕ' are *companions* if $\phi v = \phi' v$ for every v bounded and continuous on F, which here is the whole of $[-\infty, \infty)$. This concept admits of wide topological generalization, and much of this section can be correspondingly generalized. It will soon be clear that, at least for the purposes of this chapter, a pair of companions really are very much alike. On this account, Lemma 2 below reduces the general situation nearly to that for conventional gambles.

LEMMA 1. The not partially remote lotteries, ϕ and ϕ', are companions if and only if $\phi(f - \epsilon, \infty) \geq \phi'(f, \infty)$ and $\phi'(f - \epsilon, \infty) \geq \phi(f, \infty)$ for all f in $(-\infty, \infty)$ and all positive ϵ.

Proof: Let v be any continuous function bounded between 0 and 1, with $v(g) = 1$ for $g \geq f$ and $v(g) = 0$ for $g \leq (f - \epsilon)$. If ϕ and ϕ' are companions, $\phi(f - \epsilon, \infty) \geq \phi v = \phi' v \geq \phi'(f, \infty)$. This proves the "only if" clause.

If $\phi(f - \epsilon, \infty) \geq \phi'(f, \infty)$ and $\phi'(f - \epsilon, \infty) \geq \phi(f, \infty)$ for all f in $(-\infty, \infty)$ and all positive ϵ, it is easy to see that $\phi v = \phi' v$ for every bounded, continuous, monotonic v. Therefore, $\phi v = \phi' v$ if v is continuous and of bounded variation on $[-\infty, \infty)$ or is the uniform limit of such functions. If v is any function bounded and continuous on $[-\infty, \infty)$, let $v_c(f) = v(f)$ for $f \leq c$ and $v_c(f) = v(c)$ for $f \geq c$. Then $\phi v_c = \phi' v_c$, and since ϕ and ϕ' are not partially remote, $\lim \phi v_c = \phi v$ and $\lim \phi' v_c = \phi' v$. ◆

LEMMA 2. Every not partially remote ϕ has one and only one conventional companion ϕ^*, and ϕ^* is characterized by the condition that

$$\phi^*(f, \infty) = \sup_{g>f} \phi(g, \infty). \tag{1}$$

Proof: Plainly (1) does define a conventional ϕ^*. In view of (1), $\phi^*(f - \epsilon, \infty) \geq \phi(f, \infty)$;

$$\phi(f - \epsilon, \infty) \geq \phi^*(f - \epsilon, \infty) \geq \phi^*(f, \infty);$$

so Lemma 1 applies to show that ϕ and ϕ^* are companions. Lemma 1 shows also that distinct conventional lotteries cannot be companions. ◆

LEMMA 3. If ϕ and ϕ' are companions, then $\int f\, d\phi(f) = \int f\, d\phi'(f)$ if either integral is defined.

A pair of bounded, nondecreasing functions V and V' on $[-\infty, \infty)$ are *companions* if $V(f + \epsilon) \geq V'(f)$ and $V'(f + \epsilon) \geq V(f)$ for all f in $[-\infty, \infty)$ and all positive ϵ.

LEMMA 4. V and V' are companions if and only if they have the same points of continuity and agree with each other there and at $-\infty$.

LEMMA 5. If V_n approaches V and V'_n approaches V', and if V_n and V'_n are companions for each n, then V and V' are companions. If, further, V_n approaches V uniformly, then V'_n approaches V' uniformly.

Proof: To see that V and V' are companions, calculate thus for fixed f, ϵ, and δ and for sufficiently large n: $V(f + \epsilon) \geq V_n(f + \epsilon) - \delta \geq V'_n(f) - \delta \geq V'(f) - 2\delta$.

If V_n approaches V uniformly, then, in view of Lemma 4, V'_n approaches V' uniformly on the set of those fortunes that are points of continuity of V' and of every V'_n. Let G be the denumerable set of fortunes not covered by this conclusion. For any δ there are only a finite number of elements of G where the magnitude of the discontinuity of V' is as great as δ. ♦

LEMMA 6. If V and V' are companions and ϕ and ϕ' are companions, then the functions of f defined by $\int V(f + g)\, d\phi(g)$ and $\int V'(f + g)\, d\phi'(g)$ are companions.

Proof:

$$\begin{aligned}\int V(f + 2\epsilon + g)\, d\phi(g) &\geq \int V'(f + \epsilon + g)\, d\phi(g) \\ &= \int V'(f + g)\, d[\phi + \epsilon](g) \\ &\geq \int V'(f + g)\, d\phi'(g).\end{aligned}$$ ♦

Let U_n, U_ω, and U be defined for ϕ as in Section 2, and let $\hat{U}_n$ and $\hat{U}_\omega = \hat{U}$ be the corresponding functions for the conventional companion $\hat{\phi}$ of ϕ.

LEMMA 7. U_n and $\hat{U}_n$, and U_ω and $\hat{U}_\omega$, are companions. Both U_ω and $\hat{U}_\omega$ are identically 1 in $(-\infty, 0)$, or both are less than 1 there. If U_ω is less than 1 in $(-\infty, 0)$, then U_n converges uniformly to U_ω. In either case, $U_\omega = U$, and U and $\hat{U}$ are companions.

Proof: Lemma 6 and Theorem 2.2 show that U_n and $\hat{U}_n$ are companions. Lemma 5 then shows that U_ω and $\hat{U}_\omega$ are companions, which implies that both are 1 in $(-\infty, 0)$ or neither is. The assertion about uniform convergence follows from Theorem 3.1 and the last sentence of Lemma 5. The final assertion is now an application of part (*b*) of Theorem 2.15.5. ♦

Lemma 7 contains Theorems 3.1 and 3.2 without the hypothesis that ϕ is conventional. Since the central Theorem 3.3 without the hypothesis that ϕ is conventional is a corollary of Lemmas 3 and 7, the objectives of this section have been attained.

5. PROPORTIONAL STRATEGIES

Let $u(\theta)(f)$ be the utility of the θ-proportional strategy initiating at f in $[0, \infty)$, and let ϕ be the lottery on $[-\infty, \infty)$ such that

$\phi \log (A + 1) = \theta A$ for every $A \subset [-1, \infty)$. Then $u(\theta)(f)$ is the utility of the ϕ-translational strategy initiating at $\log f$. Through the final paragraph of Section 4, this observation leads to the following conclusions.

THEOREM 1. If $f = 0$, $u(\theta)(f) = 0$. If $f \geq 1$, $u(\theta)(f) = 1$. One of the following cases applies for all f in $(0, 1)$ and each can occur:

(*a*) $\theta(\epsilon, \infty) = 0$ for all positive ϵ, and $u(\theta)(f) = 0$.

(*b*) $\theta(\epsilon, \infty) > 0$ for some positive ϵ, $\int \log (1 + x)\, d\theta(x) < 0$, and $0 < u(\theta)(f) < 1$.

(*c*) $\theta(\epsilon, \infty) > 0$ for some positive ϵ, $\int \log (1 + x)\, d\theta(x) \geq 0$, $u(\theta)(f) = 1$.

(*d*) $\int \log (1 + x)\, d\theta(x)$ is undefined, and $u(\theta)(f) = 1$.

(*d′*) $\int \log (1 + x)\, d\theta(x)$ is undefined, and $0 < u(\theta)(f) < 1$.

On $[-1, \infty)$, $m(\theta) = \int x\, d\theta(x)$ is, of course, defined for all θ. If $m(\theta) \leq 0$, θ is at most fair, and cases (*a*) and (*b*) of Theorem 1 are the actual possibilities. If $m(\theta) > 0$, case (*a*) is excluded, but all the others are possible, as examples will soon show.

Thus, even if θ is superfair, the θ-proportional strategies may have probability less than 1 of attaining the goal 1 from every fortune f in $(0, 1)$. Nonetheless, for sufficiently small positive s the $[s\theta]$-proportional strategies do attain the goal with probability 1 from every positive fortune if θ is superfair. Letting

$$B(s) = \int \log (1 + x)\, d[s\theta](x) = \int \log (1 + sx)\, d\theta(x), \tag{1}$$

it will in fact be shown that $B(s) > 0$, so that case (*c*) obtains for $[s\theta]$, for all sufficiently small positive s if θ is superfair.

THEOREM 2. For every θ on $[-1, \infty)$ with $\int x\, d\theta(x) > 0$, $B(s)$ is defined for s in $[0, 1)$. It is 0 for $s = 0$. Either $B(s) = \infty$ for all s in $(0, 1)$ or $B(s) < \infty$ and is concave for s in $[0, 1)$, and $B'(0) > 0$. $B(s)$ is positive for sufficiently small s. The following cases, and only these, can occur:

1. $B(s) < \infty$ in $[0, 1]$.
   1.1. $B(s) > 0$ in $(0, 1)$.
   1.2. For a unique s_0 in $(0, 1)$, $B(s) > 0$ in $(0, s_0)$, $B(s_0) = 0$, and $B(s) < 0$ in $(s_0, 1]$.
2. $B(s) = \infty$ for all s in $(0, 1)$.
   2.1. $B(1) = \infty$.
   2.2. $B(1)$ is undefined.
      2.2.1. $u(\theta) \equiv 1$.
      2.2.2. $0 < u(\theta)(f) < 1$ for f in $(0, 1)$.

Both subcases of case 1 are compatible both with $\int x \, d\theta(x) < \infty$ and with $\int x \, d\theta(x) = \infty$, but case 2 is compatible only with $\int x \, d\theta(x) = \infty$.

Proof: Since $\log(1 + sx) \geq \log(1 - s)$, $B(s)$ is defined and is at least $\log(1 - s)$ in $[0, 1)$, and $B(0)$ is obviously 0.

Since, for s and s' in $(0, 1)$, $\log(1 + sx) - \log(1 + s'x)$ is bounded for x in $[-1, \infty)$, $B(s)$ is either infinite for all s in $(0, 1)$ or for none. If the first of these possibilities obtains, $\int_0^\infty \log(1 + x) \, d\theta(x) = \infty$, and cases 2.1 and 2.2 are the only ones possible. Since

$$sx \geq \log(1 + sx),$$

these cases cannot occur unless $\int x \, d\theta(x) = \infty$.

If $B(s)$ is finite in $[0, 1)$, then, according to the usual techniques, $B(s)$ has two derivatives in s for s in $[0, 1)$ that are computable by differentiation under the expectation sign. It is seen thus that B is concave in s for s in $[0, 1)$ and, since θ is superfair, $\dot{B}(0) > 0$. Also, $B(0) = 0$. Since, for s in $(0, 1)$, $\log(1 + x) - \log(1 + sx)$ is bounded for $x > 0$, $\int_0^\infty \log(1 + x) \, d\theta(x) < \infty$; so $B(1)$ is well defined though possibly $-\infty$. Also,

$$\int_0^\infty \log(1 + x) \, d\theta(x) = \lim_{s \to 1} \int_0^\infty \log(1 + sx) \, d\theta(x), \tag{2}$$

because $\log(1 + sx)$ approaches $\log(1 + x)$ uniformly in $x \geq 0$. By a slightly different but equally easy argument,

$$\int_{-1}^0 \log(1 + x) \, d\theta(x) = \lim_{s \to 1} \int_{-1}^0 \log(1 + sx) \, d\theta(x). \tag{3}$$

So $B(s)$ approaches $B(1)$ as s approaches 1. Putting these facts together, $B(s)$ is positive for sufficiently small positive s, and if it is not positive for all s in (0, 1) (case 1.1), it has a unique zero s_0 as required by case 1.2.

It has thus been shown that one of the five mutually exclusive cases 1.1, 1.2, 2.1, 2.2.1, or 2.2.2 must obtain. Examples will now show that these cases are all possible and will illustrate some other phenomena.

Example 1. θ is uniform on $[-1, g + 1]$ for some positive g.

$$\infty > \int x \, d\theta(x) = (g + 2)^{-1} \int_{-1}^{g+1} x \, dx = \frac{g}{2} > 0. \tag{4}$$

$B(1) = \int \log (1 + x) \, d\theta(x) = \log (g + 2) - 1$. This can be positive, 0, or negative, illustrating both parts of case 1 with $\int \theta \, d\theta(x) < \infty$.

Example 2. θ is a convex combination of the uniform distribution on $[-1, 0]$ and a distribution on $(0, \infty)$ with density $\lambda(1 + x)^{-1-\lambda}$ for some λ in (0, 1). Thus θ has density b on $[-1, 0)$ and density $\bar{b}\lambda(1 + x)^{-1-\lambda}$ on $(0, \infty)$ for some b in (0, 1).

$$B(1) = b \int_{-1}^{0} \log (1 + x) \, dx + \bar{b}\lambda \int_{0}^{\infty} (1 + x)^{-1-\lambda} \log (1 + x) \, dx < \infty. \tag{5}$$

Manipulating b for any fixed λ illustrates both parts of case 1 again, this time with $\int x \, d\theta(x) = \infty$.

Example 3. Paralleling the proof of Lemma 3.3, let θ be the distribution of $e^y - 1$ for a random variable y with $E \max (0, y) = \infty$; so $B(1) = Ey$ is ∞ or is undefined. For any such θ, case 1 does not obtain. If $B(1) = \infty$, it is case 2.1 that does. If $B(1)$ is undefined, then, according to part (*d*) of Theorem 3.4, either of the two remaining subcases can occur. ♦

THEOREM 3. If Γ is a casino with a superfair lottery θ available at 1, then Γ is superfair, $U(f) = 1$ for $f > 0$, and the $[s\theta]$-proportional family of strategies is optimal for all sufficiently small positive s.

Proof: According to Theorem 2, $B(s) = \int \log (1 + x)\, d[s\theta](x) > 0$ for sufficiently small $s > 0$; so, according to part (c) of Theorem 1, $u([s\theta])(f) = 1$ for f in $(0, 1)$. ◆

Even if every expected gain and every expected fortune f_n is infinite, the fortunes f_n can converge to 0 if you repeatedly stake too large a fraction of your fortune, as Example 2 illustrates.

An example of some interest is a superfair simple lottery, that is, a lottery under which the gambler either gains g with probability w or loses l with probability $\bar{w}$, where $wg > \bar{w}l$. Here

$$B(s) = \int \log (1 + sx)\, d\theta(x) = w \log (1 + sg) + \bar{w} \log (1 - sl),$$

and $B(1)$, of course, is $w \log (1 + g) + \bar{w} \log (1 - l)$. In a primitive casino, $l = 1$, and $B(1) = -\infty$. Therefore, in a superfair primitive casino, it is imprudent to stake a large share of your fortune repeatedly. Even if $0 < l < 1$, w and g can be so manipulated as to ensure $\int x\, d\theta(x) > 0$ and $B(1) < 0$.

It is not to be expected that the root s_0 of the equation $B(s) = 0$ can be calculated explicitly in terms of w, g, and l. However, a stake s less than s_0 can easily be found for a simple superfair θ. An elementary differentiation gives $B'(s) = wg(1 + sg)^{-1} - \bar{w}l(1 - sl)^{-1}$. Solving $B'(s) = 0$ gives $s' = (wg - \bar{w}l)/gl$. For s, the minimum of s' and 1, the $[s\theta]$-proportional strategies attain the goal with probability 1 from every $f_0 > 0$.

How does the expected number of gambles $E(N|f, s, \theta)$ required to attain the goal from a given f in $(0, 1)$ depend on s for s in $(0, s_0]$, presuming the expected number is finite? Breiman (1961) has found results along these lines. We ourselves know little about the dependence of this function on s, but it seems implausible that it should not have discontinuities if, for example, θ is two-valued and superfair. The function is not always monotonic, for according to a well-known theorem of Wald and Blackwell (Blackwell 1946; Doob 1953, p. 351), the expectation is large near $s = 0$ and, if there is an s_0, near s_0.

When $B(s) < 0$, the dependence on s of the probability $u([s\theta])(f)$ of attaining the goal 1 would also be of interest. This function can easily be discontinuous, but perhaps it must be nonincreasing in s.

11
FAIR CASINOS

1. INTRODUCTION. This chapter determines when a casino Γ is fair, that is, such that $U(f) = f$ in $(0, 1)$. It is necessary that Γ contain no superfair lottery (Section 4.4 and Theorem 10.5.3); in the presence of this condition, Γ is fair if and only if it is *at least fair*, that is, if and only if $U(f) \geq f$ for all f—or, equivalently, for some f—in $(0, 1)$.

Suppose $\Theta(1)$ contains no superfair lottery but does contain a fair nontrivial θ. If θ has bounded support, then it is easy to show that Γ is fair. But what if θ does not have bounded support? That a lottery can generate a fair casino though it has no bounded support was seen in Section 4.4. Donald Ornstein showed us, in the spring of 1958, that a fair lottery can generate an arbitrarily subfair casino. Therefore, some fair lotteries generate fair casinos, and others generate subfair casinos.

Which do which? For θ to generate a fair casino it is necessary and sufficient that

$$(1) \qquad \int_{-1}^{\infty} w \, d\theta(w) = 0, \qquad \int_{-1}^{\infty} |w| \, d\theta(w) > 0,$$

and

$$(2) \qquad \liminf_{z \to \infty} \frac{-z \int_{-1}^{z} w \, d\theta(w)}{\int_{-1}^{z} w^2 \, d\theta(w)} = 0.$$

A main purpose of this chapter is to prove this (Theorem 6.1). Another is to find necessary and sufficient conditions for any casino, not only one generated by a single lottery, to be fair (Theorem 3.1 and Theorem 3.2).

2. FAIR LOTTERIES CAN GENERATE SUBFAIR CASINOS.

This section demonstrates Donald Ornstein's conclusion that some fair lotteries generate arbitrarily subfair casinos, but by a method different from the one used by him. This conclusion will not be formally relied on later, so this section could be read only casually.

Let γ_0 be a fixed gamble and Γ_0 the smallest casino for which $\gamma_0 \in \Gamma_0(1)$. Thus, $\Gamma_0(f) = \{[fs\gamma_0 + f\bar{s}]: 0 \leq s \leq 1\}$. If Q is any nonnegative function that is fairly regular, increasing on $[0, 1]$, and equal to 1 on $[1, \infty)$, it will prove possible to choose γ_0 fair and yet such that $\gamma Q \leq Q(f)$ for all f and all γ in $\Gamma_0(f)$. According to Theorem 2.12.1, this implies that Q majorizes U; so Γ_0 is subfair if Q also satisfies $Q(f) < f$ for some f in $(0, 1)$.

Specifically, let Q restricted to $[0, 1]$ have the following properties: (*a*) $Q(f) \geq 0$ and $Q(1) = 1$; (*b*) $\dot{Q}(f) \geq \alpha > 0$ for some α; (*c*) $\ddot{Q}(f) \leq 2\alpha\beta$ for some $\beta > 0$. Such functions exist in abundance. In fact, $Q(f')$ can be arbitrarily small for any f' in $(0, 1)$. Consider, for example, $Q(f) = \epsilon^{\bar{f}/\bar{f}'}$ for f in $[0, 1]$ with $0 < \epsilon < 1$.

As one preliminary to deriving an upper bound for γQ, note that, for f and $f + g$ in $[0, 1]$,

$$Q(f + g) \leq Q(f) + g\dot{Q}(f) + g^2\alpha\beta. \tag{1}$$

Also, since $\dot{Q}(1) < \infty$, $\bar{Q}(f)/\bar{f}$ is a bounded function of f; that is, for all f in $[0, 1]$,

$$\frac{\bar{Q}(f)}{\bar{f}} \leq \alpha p \tag{2}$$

for some sufficiently large p.

To prepare for computation in this and later sections, define, for each nonnegative integer a, the ath *partial moment* $m_a(z; \theta)$ (or simply $m_a(z)$) of any θ at z in $[-1, \infty)$ thus:

$$m_a(z; \theta) = \int_{w \leq z} w^a \, d\theta(w). \tag{3}$$

Note that

$$(4) \qquad m_a(sz; [s\theta]) = \int_{w \le sz} w^a \, d[s\theta](w)$$

$$= \int_{w \le z} (sw)^a \, d\theta(w) = s^a m_a(z; \theta)$$

for all positive s.

If $\gamma = [\theta + f]$ is a fair or subfair gamble available in a casino at f for some $f \in (0, 1)$, then, using (1), (3), and (2), γQ can be estimated thus:

$$(5) \quad \gamma Q = [\theta + f]Q = \int Q(f + g) \, d\theta(g)$$

$$= \int_{g \le \bar{f}} Q(f + g) \, d\theta(g) + \bar{m}_0(\bar{f})$$

$$\le m_0(\bar{f})Q(f) + m_1(\bar{f})\dot{Q}(f) + m_2(\bar{f})\alpha\beta + \bar{m}_0(\bar{f})$$

$$= Q(f) + \alpha R(f; \theta),$$

where, in view of (2) and of the nonpositivity of $m_1(\bar{f})$,

$$(6) \qquad R(f; \theta) = \alpha^{-1}\bar{m}_0(\bar{f})\bar{Q}(f) + \alpha^{-1}m_1(\bar{f})\dot{Q}(f) + m_2(\bar{f})\beta$$

$$\le p\bar{m}_0(\bar{f})\bar{f} + m_1(\bar{f}) + m_2(\bar{f})\beta$$

$$= S(f; \theta),$$

say.

If now $\gamma \in \Gamma_0(f)$ and $\gamma \ne \delta(f)$, then $\theta = [fs\theta_0]$ for some s in $(0, 1]$, where $\theta_0 = [\gamma_0 - 1]$. Using (4) and abbreviating $\bar{f}/fs$ by z,

$$(7) \quad S(f; [fs\theta_0]) = p\bar{m}_0\left(\frac{\bar{f}}{fs}; \theta_0\right)\bar{f} + fsm_1\left(\frac{\bar{f}}{fs}; \theta_0\right) + (fs)^2 m_2\left(\frac{\bar{f}}{fs}; \theta_0\right)\beta$$

$$= \{p\bar{m}_0(z)z^2 + m_1(z)z + \beta\bar{f}m_2(z)\}\bar{f}z^{-2} \le T(z; \theta_0)\bar{f}z^{-2},$$

where

$$(8) \quad T(z; \theta_0) = p\bar{m}_0(z; \theta_0)z^2 + m_1(z; \theta_0)z + \beta m_2(z; \theta_0) \min(1, z).$$

In summary, if θ_0 is such that $T(z; \theta_0) \leq 0$ for all nonnegative z, then the U of Γ_0 is majorized by Q. It remains to construct a fair θ_0 on $[-1, \infty)$ that satisfies this inequality for all $z \geq 0$. This will be done by choosing a suitable member of the family of probability densities,

$$(9) \qquad \rho(w|\delta, b) = \begin{cases} b(1+\delta)\delta^{-1}(1+w)^{1/\delta} & \text{for } -1 \leq w \leq 0, \\ \bar{b}(1+\delta)w^{-2-\delta} & \text{for } \quad 1 \leq w, \\ 0 & \text{elsewhere,} \end{cases}$$

with $0 < b < 1$ and $\delta > 0$. As is easily checked, ρ is a probability density of mean

$$(10) \qquad \frac{-\delta}{1+2\delta}\, b + \frac{1+\delta}{\delta}\, \bar{b}.$$

Let b henceforth be so chosen that (10) vanishes, which will ensure $m_1(\theta_0) = 0$. It will be enough to remember that $b/\bar{b} = O(\delta^{-2})$ for small δ. If θ_0 is the distribution corresponding to $\rho(w|\delta, b)$, this table shows the computation of $T(z; \theta_0)$.

	$0 \leq z \leq 1$	$1 \leq z$
$p\bar{m}_0(z)z^2$	$\bar{b}z^2p$	$\bar{b}z^{1-\delta}p$
$m_1(z)z$	$\frac{-\bar{b}(1+\delta)z}{\delta}$	$\frac{-\bar{b}(1+\delta)z^{1-\delta}}{\delta}$
$\beta m_2(z)\min(1, z)$	$\frac{2\bar{b}\delta^2 z\beta}{(1+3\delta)(1+2\delta)}$	$\frac{2b\delta^2\beta}{(1+3\delta)(1+2\delta)} + \frac{\bar{b}(1+\delta)(z^{1-\delta}-1)\beta}{1-\delta}$
$T(z, \theta_0)$	$\bar{b}z[pz - \delta^{-1} + \beta O(1)]$	$\bar{b}z^{1-\delta}[p - \delta^{-1} + \beta O(1)]$

Clearly, for sufficiently small δ, $T(z; \theta_0)$ is nonpositive for all $z \geq 0$. There is great latitude in the choice of a θ_0 that makes (8) nonpositive; we have made a simple choice showing that γ_0 can have a density and attach positive probability to all neighborhoods of -1.

The section is summarized by a theorem.

THEOREM 1. For every positive ϵ and for f in $(0, 1)$, there is a rich man's casino Γ for which $U(f) \leq \epsilon$ and such that $\Gamma(1)$ contains a fair, conventional gamble other than $\delta(1)$.

3. WHEN IS A CASINO FAIR?

A casino that contains superfair gambles has been shown to be superfair. With this case out of the way, the fundamental question concerning fair casinos is answered in terms of the set of lotteries $\Theta(1) = [\Gamma(1) - 1]$ by Theorem 2 and, in a preliminary way, by Theorem 1. (As usual, $w \wedge z$ is the minimum of the two numbers w and z.)

THEOREM 1. For a casino Γ that contains no superfair lotteries, these three conditions are equivalent:

(*a*) Γ is fair.

(*b*) There is a positive z such that, for every positive ϵ, there is a θ in $\Theta(1)$ with

$$-z \int (w \wedge z)\, d\theta(w) < \epsilon \int (w \wedge z)^2\, d\theta(w). \tag{1}$$

(*c*) For every positive z and ϵ, (1) has a solution in $\Theta(1)$.

The content of Theorem 1 would plainly be unchanged by the removal of "z" before the first integral of (1). But this "z" serves to make ϵ dimensionless.

Proof that (*b*) *and* (*c*) *are equivalent:* Clearly, (*c*) implies (*b*). The reverse implication will develop from the next few paragraphs, which are of some interest in themselves.

If ϵ, z, and θ satisfy (1), then so do ϵ, sz, and $[s\theta]$. Thus, if (1) can be solved in $\Theta(1)$ for a specified positive ϵ and z, it can also be solved for that ϵ and all smaller z.

If ϵ, z, and θ satisfy (1), and $z' \geq z > 0$, let $\epsilon' = \epsilon z'/z$, and note that ϵ', z', and θ also satisfy (1).

It is now evident that, if some positive z satisfies condition (*b*) of the theorem, all positive z do; so (*b*) does imply (*c*). Still more, if some positive z fails to satisfy condition (*b*) for some positive ϵ, then all larger values of z fail for that ϵ, and all smaller z, say sz, fail for $s\epsilon$.

The next two lemmas are used in the proof that (*a*) implies (*b*).

LEMMA 1. Suppose that, for some ϵ in $(0, 1)$, every lottery θ available in a house Γ on the nonnegative reals is permitted by the quadratic function Q, $Q(x) = \epsilon x^2 + \bar{\epsilon} x$, that is, $\int Q(x)\, d\theta(x) \leq 0$. Then Q is excessive for Γ. Consequently, for any u, if $Q \geq u$, then $Q \geq U$. If it is supposed only that every θ available at every f in $[0, 1]$ is permitted by Q, then $Q \wedge 1$ is excessive and $Q \geq U$ for the usual u.

Proof: For θ permitted by Q, and $\gamma = [f + \theta]$, compute thus:

$$
\begin{aligned}
\gamma Q &= \int Q(f + x)\, d\theta(x) \\
&= \int \{Q(f) + 2\epsilon f x + Q(x)\}\, d\theta(x) \\
&= Q(f) + 2\epsilon f \int x\, d\theta(x) + \int Q(x)\, d\theta(x) \\
&\leq Q(f) + 0 + 0 \\
&= Q(f).
\end{aligned} \tag{2}
$$

Therefore, if every available θ is permitted by Q, Q is excessive; if every θ available at every f in $[0, 1]$ is permitted by Q, then $Q \wedge 1$ is excessive. In either event, Theorem 2.12.1 applies. ◆

(Lemma 1 is weaker, but also considerably simpler, than Theorem 9.4.1.)

For each measure θ and real number z, the *curtailment* $\theta \wedge z$ of θ *at* z is the measure that agrees with θ on subsets of $(-\infty, z)$ and that assigns to $\{z\}$ the value $\theta[z, \infty]$. The *curtailment* $\Gamma \wedge z$ of a house Γ based on a set of real numbers is defined by

$$(\Gamma \wedge z)(f) = \{\gamma \wedge z\colon \text{for } \gamma \in \Gamma(f)\}.$$

LEMMA 2. If Γ_1 is the curtailment at 1 of a house Γ on a set of real numbers and if u is the indicator of $[1, \infty)$, then $U_1 = U$.

Proof that condition (*a*) *of Theorem* 1 *implies* (*b*): If Γ is fair, then, according to Lemma 2, so is $\Gamma \wedge 1$. Lemma 1 then provides for each

$\epsilon > 0$ a θ^* available in $\Gamma \wedge 1$ at some f in $[0, 1]$, for which

$$
\begin{aligned}
(3) \qquad 0 &< \int Q(w)\, d\theta^*(w) = \int Q(w)\, d[\theta \wedge (1 - f)](w) \\
&\leq \int Q(w)\, d[\theta \wedge 1](w) = \int Q(w \wedge 1)\, d\theta(w) \\
&= \int \{\epsilon(w \wedge 1)^2 + \bar{\epsilon}(w \wedge 1)\}\, d\theta(w) \\
&\leq \epsilon \int (w \wedge 1)^2\, d\theta(w) + \int (w \wedge 1)\, d\theta(w),
\end{aligned}
$$

where θ is available in Γ at f and hence at 1. ◆

Proof that condition (c) of Theorem 1 implies (a): The program is to show that the utility function $f \wedge 1$ is nearly attained by a stationary family of strategies that resembles the proportional family studied in Chapter 10. For every lottery θ carried on $[-1, \infty)$ and for every z and α, where $z > \alpha > 0$, the *standard* stationary family of strategies $\sigma(f; \theta, z, \alpha)$ is defined for each f by the gamble

$$
(4) \qquad \gamma(f; \theta, z, \alpha) = \begin{cases} \left[\dfrac{\bar{f} + \alpha}{z}\, \theta + f\right] & \text{for } \dfrac{1 + \alpha}{1 + z} \leq f < 1, \\ \delta(f) & \text{elsewhere.} \end{cases}
$$

The abbreviations $\sigma(f)$ and $\gamma(f)$ will be used for $\sigma(f; \theta, z, \alpha)$ and $\gamma(f; \theta, z, \alpha)$. The idea behind the definition of standard strategy is this: If the gambler moves at all under $\sigma(f)$, he chooses the fraction fs of the lottery θ that will make $\gamma(f)[1 + \alpha, \infty) = [fs\theta + f][1 + \alpha, \infty) = \theta[z, \infty)$; he stands pat with $\delta(f)$ if the goal of 1 is already attained, or if the requisite s, namely, $(\bar{f} + \alpha)/fz$, would exceed 1, since, for $s > 1$, $[fs\theta + f]$ is not generally in $\Gamma(f)$. If $[\theta + 1] \in \Gamma(1)$, then $\gamma(f) \in \Gamma(f)$ and $\sigma(f)$ is available in Γ at f. It is helpful to think of α as small and z as large.

Denote the $\sigma(f)$-probabilities that $f_n < (1 + \alpha)/(1 + z)$ and $f_n \geq 1$ by A_n and B_n, respectively. Of course, A_n and B_n are nondecreasing in n. It will now be shown that, unless θ is trivial, $A_n + B_n$ approaches 1 for large n. Let $\beta = \theta[z, \infty)$. Since each essential gamble used by $\sigma(f)$ has probability β of leading to a fortune of at least $1 + \alpha$, $A_n + B_n \geq 1 - (1 - \beta)^n$. This proves the required point if

$\beta > 0$. Otherwise, for every n, $0 \leq f_n < 1 + \alpha$ almost certainly, and $\sigma(f; \theta, z, \alpha)$ is very like a proportional strategy in which $(1 + \alpha)$ plays the role of 0, fortunes are measured down from $1 + \alpha$, and the goal is $(1 + \alpha)/(1 + z)$. The only difference is that $\gamma(f) = \delta(f)$ between 1 and $1 + \alpha$. Formally, consider the proportional strategy p based on the lottery $[-\theta/z]$, beginning with the fortune $\bar{f} + \alpha$. It is easy to see that $A_n + B_n$ is the probability under p that some fortune among $(\bar{f} + \alpha), (\bar{f}_1 + \alpha), \cdots, (\bar{f}_n + \alpha)$ falls outside of $(\alpha, (1 + \alpha)z/(1 + z)]$. This approaches 1 in n if θ is nontrivial, as can be seen by applying Lemma 10.3.1 modified in the spirit of Section 10.4 so as to apply to general lotteries.

Let Q be any strictly convex, increasing, quadratic function of f such that $Q(f) < f$ in $[0, 1]$ and $Q(1) > 0$. As will be shown in the next paragraph, there is a θ in $\Theta(1)$, and a z and an α such that

$$Q(f) < 0 \qquad \text{for } f < \frac{1 + \alpha}{1 + z}; \tag{5}$$

$$\gamma(f; \theta, z, \alpha)(Q \wedge 1) \geq (Q(f) \wedge 1) \qquad \text{for all } f. \tag{6}$$

Consider the implications of this anticipated conclusion. First, Lemma 2.12.1 can be applied to $1 - (Q \wedge 1)$, and to the gambling house Γ_θ, where $\Gamma_\theta(f) = \{\gamma(f)\}$ (or $\{\gamma(f), \delta(f)\}$ if one prefers not to depart from the usual assumption that $\delta(f)$ is always available). Hence,

$$\sigma(f; \theta, z, \alpha)(Q(f_t) \wedge 1) \geq (Q(f) \wedge 1) \tag{7}$$

for every stop rule t. Next, if $t \geq n$, $Q(f_t) < 0$ with $\sigma(f)$-probability at least A_n; so the left side of (7) is at most $1 - A_n$. But $A_n + B_n$ approaches 1, so $\lim B_n \geq Q(f)$ for f in $[0, 1]$, whence also $u(\sigma)(f) \geq Q(f)$ in $(0, 1)$. (This argument is foreshadowed in Theorem 2.12.4.) Finally, since $Q(f)$ can be arbitrarily close to f in $[0, 1]$, the promised inequalities (5) and (6) will indeed complete the proof.

Choose α such that $Q(1 + \alpha) = 1$ and z such that

$$Q\left(\frac{1 + \alpha}{1 + z}\right) = 0,$$

as is compatible with the assumptions about Q and the requirement that $z > \alpha > 0$. This choice ensures (5). For the interesting range of f, namely, $[(1 + \alpha)/(1 + z), 1)$, and for any θ,

$$\gamma(f)(Q \wedge 1) = \int Q\left(f + \frac{\bar{f} + \alpha}{z} w\right) d[\theta \wedge z](w) \tag{8}$$

$$= Q(f) + \left\{\frac{\bar{f} + \alpha}{z} \dot{Q}(f) \int w \, d[\theta \wedge z](w) + \frac{1}{2}\left(\frac{\bar{f} + \alpha}{z}\right)^2 \ddot{Q}(f) \int w^2 \, d[\theta \wedge z](w)\right\}.$$

The coefficient of $\int w \, d[\theta \wedge z](w)$ is bounded from above, and the coefficient of $\int w^2 \, d[\theta \wedge z](w)$ is positive and bounded from below for f in the interesting range. In view of hypothesis (*c*), θ can therefore be so chosen that $\{\quad\} > 0$ for all f of interest. ◆

Theorem 1 remains true if the curtailed moments to which it refers are replaced by partial moments.

THEOREM 2. For a casino Γ that contains no superfair lotteries these three conditions are equivalent:

(*a*) Γ is fair.

(*b*′) There is a positive z such that, for every positive ϵ, there is a θ in $\Theta(1)$ with

$$-z \int_{w \leq z} w \, d\theta(w) < \epsilon \int_{w \leq z} w^2 \, d\theta(w). \tag{9}$$

(*c*′) For every positive z and ϵ, (9) has a solution in $\Theta(1)$.

(As in Theorem 1, the "z" before the first integral is to make ϵ dimensionless.)

Proof that (*b*′) *and* (*c*′) *are equivalent:* The argument that (*b*) and (*c*) are equivalent applies verbatim. ◆

Proof that (*b*′) *implies* (*b*)*:* The left side of (1) is at most the left side of (9), and the right side of (9) is at most the right side of (1). ◆

The following lemma, of possible analytical interest, will be used to show that (*b*) implies (*b*′):

LEMMA 3. Suppose that ϵ and z' are positive, and θ is any not partially remote, probability measure on $[-1, \infty]$. If

$$-z \int_{w \leq z} w \, d\theta(w) \geq 2\epsilon \int_{w \leq z} w^2 \, d\theta(w) \tag{10}$$

for all z in (z', ∞), then for all z in (z', ∞)

$$-\int_{w \leq z} w \, d\theta(w) \geq (1 + \epsilon) z \theta(z, \infty), \tag{11}$$

and

$$-z \int (w \wedge z) \, d\theta(w) \geq \frac{2\epsilon^2}{1 + 3\epsilon} \int (w \wedge z)^2 \, d\theta(w). \tag{12}$$

Proof: There is no loss in generality in assuming that θ is conventional. To see this, note that the left side of (10) can never increase rapidly, and the right side is nondecreasing. If (10) is ever violated in (z', ∞), it must then be violated throughout some nontrivial subinterval. Therefore, (10) obtains throughout (z', ∞) if and only if it obtains at the points of continuity of the functions on the right-hand and left-hand sides of (10), which are functions of bounded variation. At these points of continuity the values of the two functions are unchanged if θ', the conventional companion of θ, is substituted for θ; so (10) is satisfied throughout (z', ∞) for θ if and only if it is satisfied for θ'. A similar argument applies to (11) and (12). Specialization to conventional lotteries is now justified.

The proof involves several integrations by parts, which can be justified by, for example, (Widder 1941, Theorem 4*b*, p. 7), (Saks 1932, Theorem 14.1, p. 102), or (Hewitt 1960, Theorem A). Some instances of the general formula

$$\int_{a-0}^{b+0} \phi(t) \, d \int_{a-0}^{t} \psi(u) \, dF(u) = \int_{a-0}^{b+0} \phi(t)\psi(t) \, dF(t), \tag{13}$$

where ϕ and ψ are bounded and Borel measurable and F is of bounded variation and continuous from the right, are also involved.

Calculate thus for $z > 0$, in the notation of Section 2.

$$\bar{m}_0(z) = \int_{z+0}^{\infty} w^{-1}\, dm_1(w) \tag{14}$$

$$= -z^{-1}m_1(z) + \int_{z+0}^{\infty} w^{-2}m_1(w)\, dw.$$

$$m_2(z) = \int_{-1-0}^{z+0} w\, dm_1(w) \tag{15}$$

$$= zm_1(z) - \int_{-1-0}^{z+0} m_1(w)\, dw \geq 0.$$

If (10) obtains in (z', ∞), then, according to (15),

$$-(1 + 2\epsilon)m_1(w) \geq -\, 2\epsilon w^{-1} \int_{-1-0}^{w+0} m_1(v)\, dv \tag{16}$$

for w in (z', ∞). Therefore,

$$(1 + 2\epsilon) \int_{z+0}^{\infty} w^{-2}m_1(w)\, dw \tag{17}$$

$$\leq 2\epsilon \int_{z+0}^{\infty} w^{-3} \int_{-1-0}^{w+0} m_1(v)\, dv\, dw$$

$$= \epsilon z^{-2} \int_{-1-0}^{z+0} m_1(w)\, dw + \epsilon \int_{z+0}^{\infty} w^{-2}m_1(w)\, dw$$

for z in (z', ∞). Therefore,

$$(1 + \epsilon) \int_{z+0}^{\infty} w^{-2}m_1(w)\, dw \leq \epsilon z^{-2} \int_{-1-0}^{z+0} m_1(w)\, dw \tag{18}$$

for z in (z', ∞). Now, using (14), (18), and (15) in that order,

$$(1 + \epsilon)z^2\bar{m}_0(z) \leq -(1 + \epsilon)zm_1(z) + \epsilon \int_{-1-0}^{z+0} m_1(w)\, dw \tag{19}$$

$$= -zm_1(z) - \epsilon m_2(z) \leq -\, zm_1(z),$$

so (11) obtains for z in (z', ∞).

To see that (10) and (11) imply (12), add $\epsilon/(1 + 3\epsilon)$ times (10) to $(1 + 2\epsilon)z/(1 + 3\epsilon)$ times (11). ♦

Proof that (b) implies (b′): If (9) fails for some ϵ and z', it fails for that ϵ and all $z > z'$, and (12) applies to show that (1) fails for $\epsilon' = \epsilon^2/(2 + 3\epsilon)$ and for all $z > z'$. ◆

The proof of Theorem 2 is now complete.

Lemma 3 is not true in reverse:

Example 1. Suppose that θ is fair and conventional, with density $Ax^{-(t+2)}$ for z in $(1, \infty)$, where t is an arbitrary number larger than 1 and A is some positive number. Then,

$$\bar{m}_0(z) = \theta(z, \infty) = A \int_z^\infty w^{-(t+2)}\, dw = \frac{Az^{-(t+1)}}{t+1}, \tag{20}$$

$$-m_1(z) = A \int_z^\infty w^{-(t+1)}\, dw = \frac{Az^{-t}}{t} = \frac{z\bar{m}_0(z)(t+1)}{t} = (1 + \epsilon)z\bar{m}_0(z), \tag{21}$$

where $\epsilon = 1/t$. Therefore, (11) is satisfied. That (10) is not satisfied is easily seen thus:

$$-zm_1(z) = z \int_z^\infty w\, d\theta(w) = \int_z^\infty zw\, d\theta(w) \leq \int_z^\infty w^2\, d\theta(w), \tag{22}$$

which converges to 0 as z approaches infinity; but $m_2(z)$ approaches a finite positive limit.

The following sufficient condition for a casino to be at least fair is an immediate consequence of Theorem 2 and Lemma 3.

THEOREM 3. A casino Γ is at least fair if, for some positive z and for every positive ϵ, there is a θ in $\Theta(1)$ for which

$$- \int_{w \leq z} w\, d\theta(w) < (1 + \epsilon)\, z\theta(z, \infty). \tag{23}$$

If (23) is satisfied by z, ϵ, and θ, it is also satisfied by sz, ϵ, and $[s\theta]$; therefore if some z satisfies the condition of Theorem 3, so does any smaller z, though some z may be too large to satisfy it.

Example 2. Let Γ be generated by a single fair θ carried on $[-1, r]$, with r the least such number. The theorem applies with z any number less than r, but it does not apply with any z as large as r, since the right side of (23) vanishes for such z.

It is easy to prove Theorem 3 directly, along the lines of the proof that condition (*c*) of Theorem 1 implies condition (*a*), with the simplification that linear functions can play the role here that quadratic ones do there. This was almost done in Section 4.4.

We mention Theorem 3 because (23) is simpler than (9). As will be apparent later, however, the sufficient condition of Theorem 3 is vastly more restrictive than conditions (b') and (c') of Theorem 2.

4. FAIR, SIMPLE-LOTTERY CASINOS

To apply and illustrate the general conclusions of the preceding section, specialize $\Theta(1)$ in this section to consist of a set of simple lotteries. A *simple lottery* θ is characterized by the triple (s, t, w); θ loses the stake s with probability $\bar{w}$ and wins a prize t with probability w. The immediate object is to determine which sets of triples correspond to fair casinos. Denote the first and second moments of θ simply by m_1 and m_2, and introduce a subfairness coefficient ϕ, $\phi = -m_1/m_2$. Written out,

$$m_1 = tw - s\bar{w}, \tag{1}$$

$$m_2 = t^2w + s^2\bar{w} = (t + s)^2 w\bar{w} + m_1^2, \tag{2}$$

$$\phi = \frac{s\bar{w} - tw}{s^2\bar{w} + t^2w}. \tag{3}$$

To set aside superfair lotteries as uninteresting here, assume throughout this section that $m_1 \leq 0$ for all θ in $\Theta(1)$.

THEOREM 1. If $\Theta(1)$ consists of simple lotteries none of which are superfair, these four conditions are equivalent:

(*a*) The casino is fair.

(*b*) For every positive z and ϵ there is a θ in $\Theta(1)$ such that $t < z$ and $\phi z < \epsilon$.

(*c*) For every positive ϵ there is a $\theta \in \Theta(1)$ such that $t < \epsilon$ and $\phi < \epsilon$.

(*d*) For every positive ϵ there is a θ in $\Theta(1)$ such that $\phi < \epsilon$ and $\phi t < \epsilon$.

Proof: According to Theorem 3.2, if the casino is fair, there is, for every positive z and ϵ, a θ in $\Theta(1)$ such that $-zm_1(z;\theta) < \epsilon m_2(z;\theta)$. This inequality cannot obtain if $\epsilon \leq z \leq t$, for it would mean $zs\bar{w} < \epsilon s^2\bar{w}$, or $z < \epsilon s$. Thus, for $\epsilon \leq z$ there must be a θ with $t \leq z$ and $-zm_1 < \epsilon m_2$. And the inequality $t \leq z$ can be made strict by contracting θ slightly. Since the condition $t < z$ and $-zm_1 < \epsilon m_2$ can be satisfied for all positive z and small positive ϵ, it can be satisfied for all positive z and ϵ; so (a) implies (b).

(b) implies that, for a certain θ, $-zm_1(z;\theta) < \epsilon m_2(z;\theta)$. So, according to Theorem 3.2, (b) also implies (a).

If (b) obtains, substitute ϵ for z and ϵ^2 for ϵ in (b) to conclude (c). Conversely, if (c) obtains, choose the ϵ in (c) smaller than both the z and ϵ/z of (b). Thus (b) and (c) are equivalent.

It is evident that (c) implies (d). Finally, it will be shown that, under (d), there are θ's with t and ϕ arbitrarily small, as required by (c). For if (t, ϕ) corresponds to a θ in $\Theta(1)$, then so do all points (t', ϕ') on the semihyperbola $0 < t' \leq t$, $t'\phi' = t\phi$. Specifically, if there is a θ with $\phi < \epsilon$ and $\phi t < \epsilon$ for some ϵ in $(0, 1)$, there is also a θ' with $t' \leq \epsilon^{1/2}$ and $\phi' \leq \epsilon^{1/2}$. First, if $t \leq \epsilon^{1/2}$, let $t' = t$ and $\phi' = \phi$. Otherwise, $\theta' = [\epsilon^{1/2}\theta/t]$ is in $\Theta(1)$ along with θ. But $t' = (\epsilon^{1/2}/t)t = \epsilon^{1/2}$, and $\phi = (\epsilon^{1/2}/t)^{-1}\phi = \phi t\epsilon^{-1/2} < \epsilon^{1/2}$. ◆

COROLLARY 1. If t is bounded, the casino is fair if and only if there are θ's with ϕ arbitrarily small.

If, for every bounded lottery θ, t is interpreted as the essential upper bound of θ, the four conditions of Theorem 1 are meaningful for any casino all of whose lotteries are bounded. The proofs given in Theorem 1 that conditions (b), (c), and (d) are equivalent and imply condition (a) apply to any casino whose lotteries have bounded carriers, with almost no change. However, condition (a) does not imply condition (b) for all such casinos.

Example 1. For $0 < \delta < 1/3$, let θ_δ attach probability $\frac{1}{2} + \delta$ to -1, $\frac{1}{2} - \delta - \delta^2$ to 1, and δ^2 to δ^{-1}. Let Γ be the casino for which $\Theta(f) = \{[a\theta_\delta]: 0 \leq a \leq f, 0 < \delta < 1/3\}$. As is easily verified, Γ is fair but does not satisfy condition (b) of Theorem 1.

For some questions it is convenient to introduce the *shape param-*

eters $\hat{t}$, $\hat{\phi}$ of a lottery, where

$$\hat{t} = \frac{t}{s}; \qquad t = s\hat{t} \tag{4}$$

$$\hat{\phi} = s\phi = \frac{\bar{w} - \hat{t}w}{\bar{w} + \hat{t}^2 w}; \qquad \phi = \frac{\hat{\phi}}{s}. \tag{5}$$

These are "shape parameters" in that the $\hat{t}$ and $\hat{\phi}$ of a lottery θ are equal, respectively, to the $\hat{t}$ and $\hat{\phi}$ of $[a\theta]$ for any positive number a. If Θ_* is a set of simple lotteries, all with $s \leq 1$, then there is a smallest casino for which $\Theta(1) \supset \Theta_*$. For this casino,

$$\Theta(1) = \bigcup_{0 \leq a \leq 1} [a\Theta_*]. \tag{6}$$

If this smallest casino is at least fair, and only then, every casino for which $\Theta(1) \supset \Theta_*$ is at least fair. The following corollaries all refer to the smallest casino associated with Θ_*, called *the casino of* Θ_*, and it is to be understood that Θ_* includes no superfair gambles.

COROLLARY 2. The casino of Θ_* is fair if and only if, for every positive ϵ, $\phi < \epsilon$ and $\hat{\phi}\hat{t} < \epsilon$ for some θ in Θ_*.

COROLLARY 3. If t is bounded on Θ_*, the casino of Θ_* is fair if and only if, for every positive ϵ, $\phi < \epsilon$ for some θ in Θ_*.

COROLLARY 4. If s is bounded away from 0 on Θ_*, the casino of Θ_* is fair if and only if, for every positive ϵ, $\hat{\phi} < \epsilon$ and $\hat{\phi}\hat{t} < \epsilon$ for some θ in Θ_*.

COROLLARY 5. If t is bounded and s is bounded away from 0 on Θ_*, the casino of Θ_* is fair if, and only if, for every positive ϵ, $\hat{\phi} < \epsilon$ for some θ in Θ_*.

Example 2. Let Θ_* consist of simple lotteries θ_δ for which $s = -1$, $t = \bar{w} = \delta$. For θ_δ, $\hat{\phi} = \phi = (\delta - \delta\bar{\delta})/(\delta + \delta^2\bar{\delta}) = \delta/(1 + \delta\bar{\delta})$; so the casino of Θ_* is fair according to Corollary 5. On the other hand, as is easy to verify, the condition of Theorem

3.3 is not satisfied for this uniformly bounded casino. This suggests how far the sufficient condition for fairness stated by that theorem is from being necessary.

5. FAIR, ONE-GAME CASINOS

Recall from Section 4.7 that a casino is a one-game casino generated by θ if $\theta \in \Theta(1)$ and every element of $\Theta(1)$ is of the form $[s\theta]$ for some nonnegative s. The definition was so phrased that the set of s for which $[s\theta] \in \Theta(1)$ can be a half-open or a closed interval with 0 as its lower endpoint.

THEOREM 1. If Γ is a one-game casino generated by a θ for which $\theta(z, \infty)$ approaches 0 as z approaches ∞, these three conditions are equivalent:

(*a*) The casino Γ is fair.

(*b*) θ is nontrivial, and

$$\liminf_{z \to \infty} \frac{-z \int_{w \le z} w \, d\theta(w)}{\int_{w \le z} w^2 \, d\theta(w)} = 0. \tag{1}$$

(*c*) θ is nontrivial, satisfies (1), and is a fair lottery; that is, θ has mean 0.

Proof: Suppose first that Γ is fair. Then, according to Section 4.4, θ is at most fair and is nontrivial. Consequently, $m_2(z)$ is uniformly positive for positive z, the limit inferior in (1) is at least 0, and the numbers s for which $[s\theta] \in \Theta(1)$ are bounded from above. According to Theorem 3.2, for every positive z and ϵ there must be an s with $[s\theta]$ in $\Theta(1)$ and $-zm_1(z; [s\theta]) < \epsilon m_2(z; [s\theta])$. In view of (2.4), $-zm_1(z/s; \theta) < \epsilon s m_2(z/s; \theta)$. Since s is bounded, this shows that (1) obtains, and (*a*) implies (*b*).

Calculating in the opposite direction, (*b*) implies (*a*).

Clearly, (*c*) implies (*b*). What remains to be shown is only this: If (*b*) obtains, θ is at least fair. But if θ were not at least fair, $m_1(z; \theta)$ would be uniformly negative in z and (3.15) would contradict (1). ◆

THEOREM 2. Each of the following conditions on a nontrivial lottery θ of mean 0 implies all of its successors and none of its predecessors:

(*a*) θ has bounded support.
(*b*) θ has finite variance.
(*c*) θ is such that

$$\liminf_{z \to \infty} -z \int_{w \leq z} w \, d\theta(w) \tag{2}$$

is 0.
(*d*) θ is such that (2) is finite.
(*e*) The one-game casino generated by θ is fair.

Proof: Obviously, (*a*) implies, but is not implied by, (*b*). If (*b*) obtains,

$$-z \int_{w \leq z} w \, d\theta(w) = z \int_{w > z} w \, d\theta(w) \leq \int_{w > z} w^2 \, d\theta(w), \tag{3}$$

and the final integral in (3) approaches 0 as z approaches infinity. Therefore (*b*) implies (*c*). If $-m_1(z) = A(z \log z)^{-1}$ for z in $(2, \infty)$, then (*c*) obtains but (*b*) does not.

Obviously, (*c*) implies (*d*). If $-m_1(z) = Az^{-1}$, then (*d*) obtains but (*c*) does not.

If (*d*) obtains but (*b*) does not, then (*e*) is obvious in view of Theorem 1. If (*d*) and (*b*) obtain, then so does (*c*), but (*b*) with (*c*) implies (*e*) in view of Theorem 1. So (*d*) implies (*e*).

If $-m_1(z) = Az^{-1} \log z$ for $z > 1$, then (*d*) fails but, according to Theorem 1, (*e*) obtains. ◆

According to condition (*b*) of Theorem 2, a fair θ that has a finite moment of order 2 generates a fair casino. The order 2 is critical; that is, for every r, $1 \leq r < 2$, there is a fair θ that has a finite moment of order r but generates a subfair casino.

Example 1. If θ has a density proportional to w^{-1-r} for w in $(2, \infty)$, then θ has moments of all orders up to but not including r and, by Theorem 1, is subfair if r is less than 2. If $r = 2$, θ has an infinite second moment and does not even satisfy (2), but it generates a fair casino. If $r > 2$, the second moment is plainly finite and the casino is fair.

Theorem 1 is of course a specialization of Theorem 3.2. In the same spirit, the next theorem, partially proved in Section 4.4, specializes the rather feeble Theorem 3.3.

THEOREM 3. If Γ is a one-game casino generated by a not superfair θ, then Γ satisfies the condition of Theorem 3.3 (and is therefore fair) if and only if, for every positive ϵ, there is a positive z for which

$$-\int_{w \leq z} w\, d\theta(w) < (1 + \epsilon) z \theta(z, \infty). \tag{4}$$

Applied to a fair lottery, (4) has a neat interpretation. Namely, the expectation of the prize, given that it is more than z, is less than $(1 + \epsilon)z$. But the condition (4), like that of Theorem 3.3 from which it is derived, is excessively strong. In fact, Example 3.1 shows that some very simple distributions having finite variance (or any other moment of fixed order) do not satisfy the condition of Theorem 3.

An example will now be adduced to show that "lim inf" cannot be replaced by "lim" in Theorem 1. The same example will show incidentally that the condition of Theorem 3 can apply to a θ for which even condition (*d*) of Theorem 2 fails.

Example 2. Let θ attach probability 1 to the set consisting of -1 and the integers $n!$, $n = 1, 2, \cdots$. Further, let $-m_1(n!) = A2^n/(n + 2)!$, where A is of course determined by normalization. According to (3.15),

$$\begin{aligned} (5) \quad m_2(n!) &= \int_{w \leq n!} w\, dm_1(w) \\ &= \theta\{-1\} + \theta\{1\} + \sum_{k=2}^{n} k![m_1(k!) - m_1((k-1)!)] \\ &= B + A \sum_{k=0}^{n} k! \left[\frac{2^{k-1}}{(k+1)!} - \frac{2^k}{(k+2)!}\right] \\ &= B + A \sum_{k=0}^{n} \frac{2^{k-1}k}{(k+1)(k+2)} \\ &= B + A \sum_{k=0}^{n} \left[\frac{2^k}{k+2} - \frac{2^{k-1}}{k+1}\right] \\ &= B - \frac{A}{2} + A\frac{2^n}{n+2} \sim A\frac{2^n}{n+2}. \end{aligned}$$

Therefore, $-n!m_1(n!)/m_2(n!) \sim (n+1)^{-1}$; so (1) is satisfied, and θ generates a fair casino. But, for $n > 1$,

$$\frac{-n!m_1(n!-0)}{m_2(n!-0)} = \frac{-n(n-1)!m_1((n-1)!)}{m_2((n-1)!)} \sim 1;$$

so (1) would not be true with "lim" in place of "lim inf".

According to (3.14),

$$(6) \qquad \bar{m}_0(n!) = \int_{w>n!} w^{-1}\, dm_1(w)$$
$$= \sum_{k=n+1}^{\infty} k!^{-1} A \left[\frac{2^{k-1}}{(k+1)!} - \frac{2^k}{(k+2)!}\right]$$
$$\sim \frac{-m_1(n!)}{(n+1)!}.$$

Therefore,

$$m_1(n!-0) = m_1((n-1)!) \sim n!\bar{m}_0((n-1)!) = n!\bar{m}_0(n!-0),$$

and condition (4) of Theorem 3 is satisfied, though condition (2) of Theorem 2 is not, and θ has infinite variance.

As Example 2 shows, "lim inf" in (1) cannot be replaced by "lim." It cannot be replaced by "inf" either, though in a sense it nearly can. If the denominator in (1) is bounded away from 0 for all positive z, that is, if $\theta[-1, -\epsilon) > 0$ for some $\epsilon > 0$, then the fraction in (1) approaches 0 for small z. But if the infimum of the fraction outside of some neighborhood of 0 is 0, then its lower limit as z approaches ∞ is also 0. More fully, if θ is fair and nontrivial, either $-zm_1(z)/m_2(z)$ is bounded away from 0 on every interval bounded away from both 0 and ∞, or it is identically 0 for sufficiently large z. This follows from calculation, for $z > 0$ and $t \geq 1$, thus:

$$(7) \qquad \frac{-tzm_1(tz)}{m_2(tz)} \leq \frac{-tzm_1(z)}{m_2(z)} = t\left[\frac{-zm_1(z)}{m_2(z)}\right].$$

For fair θ, (1) can be rewritten

$$(8) \qquad \liminf_{z\to\infty} \frac{z\int_{w>z} w\, d\theta(w)}{\int_{w<z} w^2\, d\theta(w)} = 0.$$

The numerator in (8) is equal to that in (1), and the denominator of (8) differs from that of (1) only at points of discontinuity of θ. Thus, in view of (7), (1) and (8) really are equivalent. Obviously, (8) implies that

$$\liminf_{z \to \infty} \frac{z^2 \int_{w>z} d\theta(w)}{\int_{w<z} w^2\, d\theta(w)} = 0, \tag{9}$$

and this is known to be a necessary and sufficient condition for the distribution θ, which is bounded from below, to belong to the domain of partial attraction of the normal law; see, for example, (Gnedenko and Kolmogorov 1954, p. 190).

Equation (8) is definitely stronger than (9), even for fair θ, according to this example.

Example 3. Let θ be a fair lottery confined to -1 and numbers of the form $z_n = 2^{n^2}/n^2$, $n = 1, 2, \cdots$, with $\theta\{z_n\} = A2^{-n^2}$, as is possible. The fraction in (9) is, at $z_n + 0$, asymptotic to 2^{-2n-1}; so θ is in the domain of partial attraction of the normal law. But the fractions in (1) and (8) approach ∞; so θ does not generate fair casinos.

If θ is fair and $\theta(z, \infty) = \bar{m}_0(z; \theta)$ is sufficiently small for large z, then θ will generate a fair casino. For in this case, θ will have finite variance. One might conjecture that, conversely, if $\theta(z, \infty)$, for a fair θ, is sufficiently large, then θ cannot generate a fair casino. Surprisingly, this conjecture is far from true. In fact, for any lottery θ of finite mean carried on $[-1, \infty)$, there is a conventional lottery ϕ of mean 0 carried on $[-1, \infty)$ that generates a fair casino and is such that $\phi(z, \infty) \geq \theta(z, \infty)$ for all z in some interval $[z_0, \infty)$. Still more, ϕ can be made to satisfy the relatively severe condition of Theorem 3, as the next theorem states.

LEMMA 1. If θ has a finite mean, there is a conventional ϕ with finite mean such that $\phi(x, \infty) \geq \theta[x, \infty)$ for all x.

Proof: For some positive y, let $F(x) = \lim_{\epsilon \to 0} \theta(-\infty, x - y + \epsilon)$. F is the distribution function of a conventional ϕ, and $\bar{F}(x) = \phi(x, \infty) = \lim_{\epsilon \to 0} \theta[x - y + \epsilon, \infty) \geq \theta[x, \infty)$. ◆

LEMMA 2. If θ is conventional and has finite mean m, there is, for each positive ϵ, a conventional ϕ such that the mean of ϕ is at most $m + \epsilon$, $\phi(x, \infty) \geq \theta(x, \infty)$ for all x, and for each positive δ there is a positive z for which

$$\int_{w>z} w \, d\phi(w) < (1 + \delta)z\phi(z, \infty). \tag{10}$$

Proof: If $\theta(x, \infty) = 0$ for some finite x, then θ itself can serve for ϕ, as is easy to see. Assume, therefore, that $\theta(x, \infty)$ is positive for all finite x. Let $\epsilon_1, \epsilon_2, \cdots$ be a sequence of numbers in (0, 1) such that

$$\sum_{j=i+1}^{\infty} \epsilon_j < \epsilon_i^2 \tag{11}$$

for all i. For instance, let $\epsilon_i = 2^{-3^i}$. Choose $a_1 > 0$ so that

$$\int_{w>a_1} w \, d\theta(w) < \epsilon_1^2. \tag{12}$$

There is a unique b_1 for which $b_1\theta(a_1, \infty) = \epsilon_1$, and $b_1 > a_1$. Choose $a_{i+1} > b_i$ such that

$$\int_{w>a_i} w \, d\theta(w) < \epsilon_i^2, \tag{13}$$

and let b_i be the number for which

$$b_i\theta(a_i, \infty) = \epsilon_i. \tag{14}$$

Since the a_i approach ∞, and

$$a_i < b_i < a_{i+1}, \tag{15}$$

the b_i approach ∞.

For each x, let $i(x)$ be the least i for which $x < b_i$. Let $T(x) = \theta(-\infty, x]$ and $G(x) = T(a_{i(x)})$. Clearly, $G(x)$ is nondecreasing and continuous on the right, and satisfies

$$(16)\qquad \int_{w>x} w\, dG(w) = \sum_{j\geq i(x)} b_j(T(a_{j+1}) - T(a_j))$$

$$= \sum_{j\geq i(x)} b_j(\theta(-\infty, a_{j+1}] - \theta(-\infty, a_j])$$

$$< \sum_{j\geq i(x)} b_j\theta(a_j, \infty) = \sum_{j\geq i(x)}^{\infty} \epsilon_j$$

$$= \epsilon_{i(x)} + \sum_{j>i(x)} \epsilon_j < \epsilon_{i(x)} + \epsilon_{i(x)}^2.$$

Let $F(x) = \min(T(x), G(x))$. Clearly, F is a distribution function, and, for some conventional ϕ, $\phi(x, \infty) = \bar{F}(x) \geq \bar{T}(x) = \theta(x, \infty)$ for all x, as required.

$$(17)\qquad \int_{w>x} w\, d\phi(w) < \int_{w>x} w(dT(w) + dG(w))$$

$$= \int_{w>x} w\, d\theta(w) + \int_{w>x} w\, dG(w)$$

$$< \int_{w>x} w\, d\theta(w) + \epsilon_{i(x)} + \epsilon_{i(x)}^2$$

$$< \epsilon_i^2 + (\epsilon_i + \epsilon_i^2) \qquad \text{if } x = b_i - 0$$

$$= \epsilon_i(1 + 2\epsilon_i) = b_i\theta(a_i, \infty)(1 + 2\epsilon_i)$$

$$= b_i\phi(a_i, \infty)(1 + 2\epsilon_i) = b_i\phi(x, \infty)(1 + 2\epsilon_i)$$

$$= x\phi(x, \infty)(1 + 2\epsilon_i + 0). \quad \blacklozenge$$

LEMMA 3. If θ is conventional and has a finite mean, there is a conventional ϕ with 0 mean such that $\phi(x, \infty), = \theta(x, \infty)$ for all sufficiently large x. If $\theta[-1, \infty) = 1$, ϕ can be so chosen that $\phi[-1, \infty) = 1$.

THEOREM 4. If θ has a finite mean, there is a conventional ϕ with mean 0 such that $\phi(x, \infty) \geq \theta[x, \infty)$ for all sufficiently large x, and such that (10) has a solution in z for each positive δ. If $\theta[-1, \infty) = 1$, ϕ can be chosen so that $\phi[-1, \infty) = 1$.

6. OPTIMAL STRATEGIES FOR FAIR CASINOS

Suppose that every $\gamma \in \Gamma(1)$ is countably additive on the Borel subsets of $[0, \infty)$. It is then natural to inquire about the existence of optimal Borel-measurable strategies, where $\sigma = \sigma_0, \cdots, \sigma_n, \cdots$ is, as usual, Borel measurable if each σ_n is countably additive on the Borel sets A, and $\sigma_n(f_1, \cdots, f_n)A$ is Borel measurable in $f_1, \cdots, f_n$ for each A.

As is not difficult to see, unless there is some $\gamma \in \Gamma(1)$ other than $\delta(1)$ that is fair, has bounded support, and attains its upper bound with positive probability, there is no optimal Borel-measurable σ. On the other hand, the presence of such a gamble in a fair casino ensures the existence of an optimal Borel-measurable strategy. The simplest one is the bold strategy based on that γ. As usual, to play boldly means to stake just enough to reach the goal when rich enough to do so and otherwise to stake all.

If a conventional fair gamble γ attains its essential upper bound with positive probability but does not attain its essential lower bound with positive probability, then a gambler who uses the bold strategy based on γ, or indeed any optimal strategy based on γ, is in the uncomfortable position of having positive probability of gambling forever. Equivalently, there is no optimal Borel-measurable wide-sense policy based on such a γ, where (σ, t) is a *wide-sense policy* if t stops with inner σ-probability 1. A necessary and sufficient condition for the existence of a Borel-measurable, optimal wide-sense policy based on a given fair, conventional γ is that both the essential least upper bound and the essential greatest lower bound of γ be attained with positive probability and that they differ from each other. Clearly, the bold strategy σ, with the associated natural incomplete stop rule t—stop when the fortune is 0 or 1—is such an optimal wide-sense policy. Thus, if a fair Γ has available such a fair γ, then measurable, optimal wide-sense policies do exist. The existence of such a fair γ is, however, not necessary for the existence of measurable, optimal wide-sense policies. It suffices that there be in $\Gamma(1)$ a fair, conventional gamble γ, other than $\delta(1)$, that attains its least upper bound with positive probability, and a fair, conventional gamble γ', other than $\delta(1)$, with bounded support, that attains its lower bound with positive probability. Indeed, this condition is both necessary and sufficient.

Return now to the question of the existence of optimal strategies when no gamble attains its least upper bound with positive probability. As was observed above, under these conditions no Borel-measurable, optimal strategies exist. Yet there are optimal strategies for at least some such fair casinos.

Example 1. Let $\Gamma(f) = [f\Theta(1) + f]$ for all f in $[0, \infty)$, where $\Theta(1)$ contains all lotteries that are fair and countably additive on the Borel sets of $[-1, \infty)$ except those that attain their upper bounds with positive probability. An optimal family of strategies, and of policies with two moves, can be constructed for this casino. Let $(0, 1)$ be partitioned into a denumerable number of sets B_n, each of inner Lebesgue measure 0. Let $\sigma_0 = \sigma_0(f)$, for f in $(0, 1)$, be any element of $\Gamma(f)$ that is carried by $(0, 1)$ and that attaches probability 0 to each of the B_n; let $\sigma_0(f) = \delta(f)$ elsewhere. (Though completely additive on Borel sets, σ_0 is not, and is not here expected to be, completely additive on all the sets for which it is defined.) Let $\sigma_1(f)$, for f in B_n, be any element of $\Gamma(f)$ that is carried by $\{0\} \cup (1 + (2n)^{-1}, 1 + n^{-1}]$; let $\sigma_1(f) = \delta(f)$ elsewhere. Let $\sigma_n(f) = \delta(f)$ for all $n \geq 2$ and all f. The strategy just defined stagnates with probability 1 but almost never at $f = 1$. With evident notation,

$$u(\sigma(f)) \geq \int \frac{f_1}{1 + n^{-1}(f_1)}\, d\sigma_0(f_1|f) \tag{1}$$

$$\geq \int_{n(f_1) \geq m}$$

$$\geq \frac{m}{m+1} \int_{n(f_1) \geq m} f_1\, d\sigma_0(f_1|f)$$

$$= \frac{m}{m+1} f.$$

So the strategy σ is worth f and is therefore optimal.

12 THE SCOPE OF GAMBLER'S PROBLEMS

1. INTRODUCTION.

Our general notion of a gambler's problem grew gradually out of our study of specific problems. In arriving at the notion, we sought a mathematically natural and convenient setting for these problems, without aiming at great generality. However, various other problems that at first seemed not to fall within the theory of gambler's problems actually do so; and this theory is closely akin to other general theories already current in mathematics. The object of this final chapter is to bring out this flexibility, generality, and kinship, by means of suggestive examples.

One respect in which the theory developed in this book is not as flexible as might be desired is that it is restricted to bounded utilities. A suggestion toward the relaxation of the restriction was made in Section 2.8. Some examples in this chapter presume utilities unbounded from one side, which is a relatively simple extension of Chapter 2.

2. FORTUNES

Much of the flexibility of the general gambler's problem derives from the diversity of the objects that can play the role of fortunes.

For the first illustration of this principle, consider so generalizing the definition of a gambling house that the set of gambles available to the gambler depends not only on his current fortune but also on his previous fortunes. (For example, there could be rules requiring the gambler to desist from certain gambles unless he has won, or lost, a specified sum of money sometime since beginning to play.) The apparent generalization can easily be made, but it is apparent only; for the general situations are seen as instances of the special ones on suitably reinterpreting "fortune".

Imagine, in fact, rules which, interpreted with respect to a set F of fortunes f, allow the gambler who has experienced the partial history $(f_0, f_1, \cdots, f_n)$—in this context, f_0 is reckoned as part of the partial history—to choose a gamble γ on F from a set $\Gamma_n(f_0, f_1, \cdots, f_n)$. Only if $\Gamma_n(f_0, f_1, \cdots, f_n) = \Gamma_0(f_n)$ for all partial histories is the situation a gambling house (in the technical sense of Section 2.4). Yet, if the partial histories $f' = (f_0, f_1, \cdots, f_n)$, rather than the elements f of F, are regarded as fortunes, the rules evidently do define a gambling house on the set F' of these fortunes. More formally, let $\Gamma'(f')$ consist of those gambles γ' on F' constructed from elements γ of $\Gamma_n(f_0, f_1, \cdots, f_n)$, thus: The probability under γ' of the set of all immediate continuations $(f_0, f_1, \cdots, f_n, f_{n+1})$ of the fortune $f' = (f_0, f_1, \cdots, f_n)$ is 1, and the probability under γ' of a subset of these continuations is the probability under γ of the corresponding set of values of f_{n+1}.

With this reinterpretation, strategies with initial fortune f in the generalized gambling house are easily seen to be isomorphic to strategies in F' with the initial fortune $f' = (f)$. To a utility u on F corresponds a utility u' on F' defined by $u'(f_0, f_1, \cdots, f_n) = u(f_n)$.

Problems can arise, in which the gambler's utility depends not only on his terminal fortune but on his initial fortune and the history of his fortunes up to termination, as when the pleasures and pains of the gambler are affected by the course of play. The artifice of regarding partial histories of fortunes in F as fortunes in F' is again applicable to reduce this extension of gambler's problems to the original form. Some problems whose solution was thus facilitated are treated in (Dubins 1962).

Suppose that the gambles available to a gambler depend not only on his present and past fortunes but also on decisions he has made in the past. For example, there might be some gamble that can be made only once or that can be made only if a deposit has been paid several moves in advance. Here again, the situation can be met either by modifying the original theory or by so reinterpreting fortune and gamble as to fit the superficially new situation into the original theory.

When $\Gamma_n(f_0, f_1, \cdots, f_n)$ depends on its arguments in certain special ways, a fortune f' in the new interpretation need not be taken to be the whole past history. Two suggestive possibilities are $f' = (f, n)$ and $f' = \max (f_0, f_1, \cdots, f_n)$.

In general, the space of fortunes F of a gambler's problem can be enlarged arbitrarily without material effect on the problem. In order to take advantage of the general theory of this book, F must be chosen large enough for the gambler's opportunities Γ and his aspirations u to be functions of f alone; simplicity often dictates an F about as small as is compatible with this condition.

3. OPINION AS A COMPONENT OF FORTUNE

The following example about a blindfolded roulette player, which at first seems outside of the original theory, introduces still other interpretations of fortune. The player deposits d dollars with a banker who, according to the player's instructions, bets s_1 dollars (where $s_1 \leq d$) on red at a roulette table. As a result, the banker will then hold for the player either $d + s_1$ or $d - s_1$, say d_1, dollars according to the probabilities w and $\bar{w}$. The player, from whom knowledge of d_1 is withheld, may now terminate play, in which case he goes home with d_1 dollars, or he may offer to bet s_2 dollars on red. If $s_2 > d_1$, he is declared bankrupt and goes home empty-handed; otherwise the bet is made for him, and the banker then holds $d_2 = d_1 \pm s_1$ dollars, and so on. The gambler's cash holdings (or their history) cannot be regarded as his fortune if his range of choice at the nth move is to be a function of his fortune; for he does not necessarily know these quantities. What he does know at time n (that is, after his nth gamble, but before his $(n + 1)$st) is d, s_1, $\cdots$, s_n, and whether he has been declared bankrupt. One might therefore let f_n be this sequence, together with the gambler's status with respect to bankruptcy. The gambler with fortune f_n can choose to terminate or to stake s_{n+1}. What utility is to be associated with the former and what gamble with the latter?

From the gambler's point of view at f_n, d_n is a random variable whose distribution he calculates from f_n (that is, from what he knows).

If he is interested in maximizing the expected value of some function u of cash, then he views termination at f_n as having the utility $E(u(d_n)|f_n)$, and this can appropriately be regarded as $u(f_n)$. Similarly, if he stakes an amount s_{n+1}, the gambler can calculate the probability that this will bankrupt him, and the gamble γ associated with the choice of s_{n+1} at f_n is confined to the two new fortunes $(d, s_1, \cdots, s_{n+1}, \textit{Solvency})$ and $(d, s_1, \cdots, s_{n+1}, \textit{Bankruptcy})$ with probabilities that depend in a calculable way on s_{n+1} and f_n.

In this problem, the details about f_n are important to the gambler only insofar as they enable him to calculate his current distribution for d_n. Instead of taking the patterns of information as fortunes, we might also have formulated the problem in terms of distributions ϕ of cash on hand. This is an idea of widespread applicability. In the present instance: the initial distribution ϕ is $\delta(d)$; the utility of ϕ regarded as a fortune is $u(\phi) = \int u(d)\, d\phi(d)$; and a stake s converts ϕ into $\delta(0)$ with probability $\phi(-\infty, s)$ and with the complementary probability $\phi[s, \infty)$ into a mixture of two distributions, weighted according to the fixed probabilities w and $\bar{w}$ of winning the next bet, where the two components of the mixture consist of the distribution ϕ truncated to $[s, \infty)$ and then shifted forward or backward by s.

With fortunes regarded as distributions of cash, it is easy to contemplate variations in the problem of the blindfolded roulette player. For example, any gambling house Γ defined on cash can be substituted for the game red-and-black, and various rules permitting the gambler to purchase information about cash on hand can be introduced. Some such rules (for instance, a rule that allows him once or twice to attempt a bet that he cannot afford with a mere warning or upon payment of a fine) would require the gambler to reckon something in addition to ϕ as part of his fortune.

4. NONLEAVABLE HOUSES

Almost all the specific gambling houses studied in this book have been leavable, that is, such that $\delta(f) \in \Gamma(f)$ for each f. But finding slight advantage, and some disadvantage, in introducing leavability as part of the definition of gambling house, we did not do so. Our definition of nonleavable gambling houses and their utilities, being little motivated by actual situations, might have been expected to have few appealing applications. It is therefore a welcome surprise that certain problems which have been studied by others, and which could be viewed as leavable problems only artificially at best, have natural interpretations as nonleavable problems. A "search problem" is described here as one illustration, and another illustration is given in Section 6.

A button is lost in one of, say, three sewing baskets, A, B, and C. A quick look into A will reveal the button with positive probability α should it be there; the corresponding probabilities for B and C are β and γ. The problem is to minimize the expected number of searches made by one who initially associates the probabilities p_0, q_0, and r_0 with the presence of the button in A, B, and C. It is known to be optimal to look into A, for instance, at a given moment if and only if the present probability αp_n of finding the button in A immediately is at least as large as the corresponding probabilities βq_n and γr_n for the other baskets. (Closely related conclusions are announced in (Bellman 1952).)

How can this problem be viewed as a gambler's problem? By now, it will seem natural to take for a generic fortune an opinion (p, q, r) and a number of past searches i. The button may be regarded as found if p, q, or r is 1, in which case the gambler is free to stagnate—and will prefer to. Otherwise, the gambler is obliged to increase i by 1 and to look into one of the baskets. Looking into A, for instance, will change (p, q, r) to $(1, 0, 0)$ if he finds the button there, which has probability αp, and to $(\bar{\alpha}p, q, r)/(\bar{\alpha}p + q + r)$ if he does not.

If the utility of the fortune $(p, q, r; i)$ is reckoned as $-i$, then $-u(\sigma, t)$ is nondecreasing in t; so $-u(\sigma) = \lim_t - u(\sigma, t)$, which is just what is meant by "the expected number of searches" in the original formulation of the problem. Incidentally, in this problem t is almost certainly bounded, as is suggested by Theorem 2.7.2, since each gamble assigns probability 1 to a finite set of fortunes. Therefore, $\lim_n u(\sigma, n) = \lim_t u(\sigma, t)$.

5. DISCOUNTING THE FUTURE

A group of problems known as two-armed bandit problems are currently attracting theoretical interest. (See, for example, (Robbins 1952), (Bellman 1956), and (Feldman 1962).) The name "two-armed bandit" was introduced by Frederick Mosteller in connection with experiments in the psychology of learning. For an illustration of the idea, consider the following situation.

A gambler is allowed without charge to play two slot machines of unknown characteristics for a total of, say, 100 plays. According to what strategy should he decide to allocate his nth play to the first or to the second machine? Assume for definiteness that the first machine when played pays a dollar with frequency p_1 and the second a dollar with frequency p_2. The gambler's opinion at the nth stage of play is given by a probability distribution ϕ_n on the (p_1, p_2)-square (beginning with an initial opinion ϕ), and ϕ_n changes from play to play according to Bayes's theorem. The problem can be formulated as a gambler's problem thus. A fortune consists of: a distribution ϕ, the number a of plays as yet available, and the amount of cash c thus far won. The utility of the fortune is simply c. For each fortune f with $a > 0$, two gambles are available. Both diminish a by 1. The first either augments c by 1 and, according to Bayes's theorem, replaces $d\phi(p_1, p_2)$ by $p_1\, d\phi(p_1, p_2)/\int p_1\, d\phi(p_1, p_2)$ or else leaves c unaltered and replaces $d\phi(p_1, p_2)$ by $(1 - p_1)\, d\phi(p_1, p_2)/\int (1 - p_1)\, d\phi(p_1, p_2)$ with probabilities $\int p_1\, d\phi(p_1, p_2)$ and $\int (1 - p_1)\, d\phi(p_1, p_2)$, respectively. The second gamble is of course analogous. When $a = 0$, play is in effect over; only $\delta(f)$ is then available.

In one variant of the two-armed bandit problem (Bellman 1956), the gambler is allowed to play indefinitely, but he discounts the future at 100ϵ per cent per play; that is, the current value to him of future payments $d_{n+1}, d_{n+2}, \cdots$ is $\Sigma_r (1 - \epsilon)^r d_{n+r}$.

Insofar as a two-armed-bandit problem is meant to suggest some aspects of a going concern such as a business or a medical research program, this formulation may represent a certain gain in realism. For a going concern does not ordinarily look forward to a predetermined last day of business, nor does it see distant gains and losses as equal in importance to immediate ones. A similar viewpoint might be appropriate to a clinic in which closely competing treatments are being tried side by side.

If, for example, the gambler is interested in expected cash and has no other reason to discount the future than the possibility that the two slot machines will simultaneously go out of order (or be turned off) with a probability of ϵ at each play, then clearly he will evaluate the future as postulated.

The problem is not now expressed in terms of the value to the gambler of his present fortune, that is, in terms of past payments to him and his present opinions about the parameters of the machines; so it is not yet a gambler's problem in the technical sense of the term. However, since the initial value to the gambler of a future sequence of payments $d_1, d_2, \cdots$ is $\Sigma (1 - \epsilon)^r d_r$, his strategic position is the same as that of a gambler who does not discount future payments but who is actually paid after the $(n - 1)$st gamble only $(1 - \epsilon)^n$ times what the slot machine emits, that is, $d'_n = (1 - \epsilon)^n d_n$. Thus the problem is to all intents and purposes a gambler's problem in which the nth fortune consists of the nth opinion about (that is, the probability distribution of) the parameters of the slot machines, together with the sum of the payments d'_r accumulated by the end of the nth play. The utility of this fortune is the accumulated payments.

6. A DIFFERENT APPROACH TO THE INDEFINITE FUTURE

A gambler knows that one of the arms of a certain two-armed bandit, the "correct" arm, returns a dollar with probability 3/4 and that the other, the "wrong" one, does so with probability 2/3. Suppose his problem is to play forever but with a strategy that minimizes the expected number of pulls on the wrong arm.

As David Blackwell pointed out to us, if the problem is taken literally, the utility depends on the entire infinite history even more conspicuously than it did in the search problem of Section 4. This problem was solved by Feldman (1962). Especially since many similar problems remain unsolved, there may be some interest in pointing out how they can be viewed as ordinary nonleavable gambler's problems.

At any given state of play, it is plainly compatible with the original formulation of the problem to reckon the gambler's fortune as consisting of what he then knows: his initial probability that the first arm was the correct one, and what arm he pulled and what happened in each trial to date. The probability $p(f)$ for the gambler with the fortune f that the first arm is the correct one and $e(f)$, the expectation for him of the number of errors to date, are calculable by the familiar rules for conditional probabilities and conditional expectations. (The gambler would seem to have no real interest in f beyond the quantities p and e, and he could actually regard them as constituting his whole fortune without any loss to himself.)

Pulling the right arm or the left one results in a random change $\gamma_R(f)$ or $\gamma_L(f)$ in the gambler's fortune, and this pair of gambles constitutes $\Gamma(f)$.

Perhaps it is obvious that letting $u(f) = -e(f)$ leads to the objective prescribed by the problem as originally formulated, but the point should be checked, thus. The total expected number of errors under σ would not be well defined except for the countably additive convention according to which it is taken to be the limit as n approaches ∞ of the expected number of errors up to time n. But the expectation under σ of the number of errors up to time n is the same as $\sigma(e(f_n))$, the expectation under σ of the expected number of errors at time n given f at time n, or $e(\sigma, n)$. So the original objective of the problem is, in effect, to maximize $-\lim_n e(\sigma, n)$; the limit exists because the expected number of errors up to time n is nondecreasing. The limit with respect to n is the same as it would be with respect to general stop rules t, because in this problem, as in the search problem of Section 4, every stop rule is bounded almost surely. Thus the original objective is indeed to maximize $-e(\sigma)$ over strategies available at the initial fortune.

7. DYNAMIC PROGRAMMING

The examples in this chapter, and many others, suggest that gambler's problems are more or less the same as problems of dynamic programming, a concept introduced by Bellman and exemplified in (Bellman 1952, 1957), (Karlin 1955), and (Howard 1961). We say "more or less" because, in our opinion, a unique definition of such general concepts as those of dynamic programming and gambler's problem cannot fruitfully be attempted. Thus, while the term "gambler's problem" is ordinarily used in this book for a definite formal concept, the term has also a looser, informal connotation. For example, there are natural problems of gambling that require continuous, as opposed to discrete, time for their formulation, and so fall outside the structure of this book. The problems of Section 9.4, for instance, invite reformulation in continuous time; for they would then have natural optimal strategies.

Dynamic-programming problems are by no means always stochastic. Nonstochastic problems of course correspond to gambling houses in which every element of each $\Gamma(f)$ is some $\delta(g)$.

8. BAYESIAN STATISTICS

Gambling problems in which the distributions of various quantities are prominent in the description of the gambler's fortune seem to embrace the whole of theoretical statistics according to one view (which might be called the decision-theoretic, Bayesian view) of the subject. Before attempting any general definition of statistical problems in this sense, we illustrate the concept by a famous and important example due mainly to Abraham Wald (whose point of view was not Bayesian).

There is a two-valued random variable H that is 1 with probability p_0 and 2 with probability $\bar{p}_0$. Conditionally on H, a sequence of random variables x_n is independently and identically distributed according to one of two distributions η_H. The gambler's object is to guess H; he receives an award A, say \$1,000, if his guess is correct. If he guesses immediately, his expected income for a reasonable guess is $\max(p_0 A, \bar{p}_0 A)$; call it $A(p_0)$. Before making his guess, however, he is permitted to see the data $x_1, x_2, \cdots$ until he is satisfied, paying an observation cost c for each x_n that he chooses to see. After each such observation, the gambler will associate a new probability p_n with $H = 1$ in accordance with Bayes's theorem. In particular, if the η_H are discrete and countably additive, as will here be assumed for definiteness and clarity, then

$$p_{n+1} = \frac{p_n \eta_1(x_{n+1})}{p_n \eta_1(x_{n+1}) + \bar{p}_n \eta_2(x_{n+1})}. \tag{1}$$

The situation can be formulated as a gambler's problem in which a fortune consists of a probability p and an amount C already paid for observations. The gambler's utility for such a fortune is $A(p) - C$. The gambles available to him are either to remain at the fortune (p, C) or to increase C by c and to exchange p for $p\eta_1(x)/[p\eta_1(x) + \bar{p}\eta_2(x)]$ with probability $[p\eta_1(x) + \bar{p}\eta_2(x)]$.

In accordance with Theorem 3.9.5, a natural stationary family of optimal strategies in this problem is to stagnate when and only when (p, C) is adequate, that is, when $U(p, C) = A(p) - C$. In this problem, it is almost obvious that $U(p, C) = U(p, 0) - C$. Consequently, the heart of the problem of finding an optimal strategy is, then, to find the set S of values of p where $U(p, 0) = A(p)$. Readers interested in the solution of this problem of testing a simple dichotomy are referred to (Wald and Wolfowitz 1948), (Arrow, Blackwell, and Girshick 1949), and (Wald 1950). Here we want only to indicate that this problem can be viewed as a gambler's problem.

From the point of view of decision-theoretic statistics, the gambler in this problem is a person who must ultimately act in one of two ways (the two guesses), one of which would be appropriate under one hypothesis ($H = 1$) and the other under its negation ($H = 2$). First, however, he may repeat a pertinent experiment (observation of the x_n) at a cost c per experiment. His behavior, if he is prudent, will be guided by his successive opinions p_n, the cost c per observation, and the award A.

Many problems, of which this one is an instance, are roughly of the following type. A person's opinion about unknown parameters is described by a probability distribution; he is allowed successively to purchase bits of information about the parameters, at prices that may depend (perhaps randomly) upon the unknown parameters themselves, until he finally chooses a terminal action for which he receives an award that depends upon the action and the parameters. (The purchases and the terminal action are, in effect, gambles.) Some of the ideas of this branch of statistics are explained in (Savage 1954, Chapter 7). Other references to Bayesian statistics are (Raiffa and Schlaifer 1961) and (Savage 1962), which has a large bibliography.

9. STOCHASTIC PROCESSES

What is the relationship of gambler's problems (in the technical sense of this book) to probability theory in general? "Strategy" is a finitely additive reinterpretation of the concept of "stochastic process in discrete time", the countably additive interpretation of which is widely adopted in probabilistic literature, as in (Doob 1950), (Loève 1960), and (Rosenblatt 1962). The analogy between the two interpretations is close and systematic. In particular, solutions to gambler's problems typically imply (and, in a sense, strengthen) solutions of corresponding problems in a countably additive framework, as is frequently illustrated in this book and in (Dubins 1962). Finitely additive, continuous-time stochastic processes and gambling remain to be explored.

Regarding strategies and discrete-time stochastic processes as different technical implementations of the same idea, a gambling house is a class of stochastic processes, and the U and V of a gambling house define sharp inequalities on such a class of stochastic processes. Thus, for example, the class of nonnegative, lower semimartingales is the universal fair casino, that is, the gambling house on the nonnegative real numbers for which $\Gamma(f)$ consists of all gambles γ with $\int f' \, d\gamma(f') \leq f$. The fact that a gambler in this casino can increase his fortune k-fold with probability k^{-1} but no more (for $k \geq 1$) is a familiar fact about nonnegative semimartingales. A solution of gambler's problems implies an inequality for stochastic processes; for the problem of Section 9.4, this implication was mentioned at the end of Section 1.2. The search for a bound on some probability associated with a stochastic process or class of processes can sometimes usefully be reformulated as a gambler's problem, as is illustrated in (Dubins 1962) and (Blackwell 1964). In short, much of the mathematical essence of a theory of gambling resides in the discovery and demonstration of sharp inequalities that pertain to interesting classes of stochastic processes.

BIBLIOGRAPHY

Addison, J. W., 1958. Separation principles in the hierarchies of classical and effective descriptive set theory. *Fundamenta Mathematicae* **46** 123–135.

Arrow, K. J., D. Blackwell, and M. A. Girshick, 1949. Bayes and minimax solutions of sequential decision problems. *Econometrica* **17** 213–243.

Bachelier, Louis, 1900. *Théorie de la Spéculation.* Paris. (Thesis.)

Bachelier, Louis, 1912. *Calcul des Probabilités.* Gauthier-Villars, Paris.

Bachelier, Louis, 1914. *Le Jeu, la Chance et le Hasard.* Flammarion, Paris. (Reprinted in 1929.)

Bachelier, Louis, 1938. *La Spéculation et le Calcul des Probabilités.* Gauthier-Villars, Paris.

Bahadur, R. R., 1958. A note on the fundamental identity of sequential analysis. *Annals of Mathematical Statistics* **29** 534–543.

Bellman, Richard, 1952. On the theory of dynamic programming. *Proceedings of the National Academy of Sciences of the United States of America* **38** 716–719.

Bellman, Richard, 1956. A problem in the sequential design of experiments. *Sankhyā* **16** 221–229.

Bellman, Richard, 1957. *Dynamic Programming.* Princeton University Press, Princeton, New Jersey.

Bernoulli, Jacob (James), 1713. *Ars conjectandi.* Thurnisius Brothers, Basel.

Bertrand, J., 1907. *Calcul des Probabilités,* second edition. Gauthier-Villars, Paris.

Blackwell, David, 1946. On an equation of Wald. *Annals of Mathematical Statistics* **17** 84–87.

Blackwell, David, 1954. On optimal systems. *Annals of Mathematical Statistics* **25** 394–397.

Blackwell, David, 1964. Probability bounds via dynamic programming. *Proceedings of Symposia in Applied Mathematics*, American Mathematical Society, Providence, Rhode Island, **XVI** 277–280.

Breiman, Leo, 1961. Optimal gambling systems for favorable games. *Proceedings of the Fourth Berkeley Symposium on Mathematical Statistics and Probability,* University of California Press, Berkeley, California, **1** 65–78.

Chung, K. L., and W. H. J. Fuchs, 1951. On the distribution of sums of random variables. *Memoirs of the American Mathematical Society* **6** 1–12.

Coolidge, J. L., 1908–1909. The gambler's ruin. *Annals of Mathematics* **10** 181–192.

de Finetti, Bruno, 1930. Sulla proprietà conglomerativa delle probabilità subordinate. *Rendiconti dell'Istituto Lombardo* **63** 414–418.

de Finetti, Bruno, 1937. La prévision: ses lois logiques, ses sources subjectives. *Annals de l'Institut Henri Poincaré* **7** 1–68.

de Finetti, Bruno, 1939. La teoria del rischio e il problema della "rovina dei giocatori". *Giornale dell'Istituto Italiano degli Attuari* **10** 41–51.

de Finetti, Bruno, 1940. Il problema dei "pieni". *Giornale dell'Istituto Italiano degli Attuari* **11** 1–88.

de Finetti, Bruno, 1949. Sull impostazione assiomatica del calcolo delle probabilità. *Annali Triestini* **19** 29–81.

de Finetti, Bruno, 1950. Aggiunta alla nota sull'assiomatica della probabilità. *Annali Triestini* **20** [Series 4, Volume 4, second section (science and engineering)] 5–22.

de Finetti, Bruno, 1955. La struttura delle distribuzione in un insieme astratto qualsiasi. *Giornale dell'Istituto Italiano degli Attuari* **18** 16–28.

de Finetti, Bruno, 1955*a*. Sulla teoria astratta della misura e dell'integrazione. *Annale di matematica pura ed applicata* (*Serie IV*) **40** 307–319.

De Moivre, Abraham, 1711. De mensura sortis. *Philosophical Transactions* **27** 213–264.

De Moivre, Abraham, 1718. *The Doctrine of Chances.* Printed for the author by W. Pearson, London.

de Rham, Georges, 1956–1957. Sur quelques courbes definies par des équations fonctionelles. *Rendiconti del Seminario Matematico dell'Università e del Politecnico di Torino* **16** 101–113.

Derman, C., and H. Robbins, 1955. The strong law of large numbers when the first moment does not exist. *Proceedings of the National Academy of Sciences of the United States of America* **41** 586–587.

Descartes, René, 1925. Livre troisième, la geometrie. *The Geometry of René Descartes.* Translated by David Eugene Smith and Marcia L. Latham. The Open Court Publishing Company, Chicago, Illinois; London.

Dickson, Leonard Eugene, 1939. *New First Course in the Theory of Equations.* John Wiley & Sons, Inc., New York.

Dieudonné, J., 1960. *Foundations of Modern Analysis.* Academic Press Inc., New York.

Doob, J. L., 1953. *Stochastic Processes.* John Wiley & Sons, Inc., New York.

Dubins, Lester E., 1962. Rises and upcrossings of nonnegative martingales. *Illinois Journal of Mathematics* **6** 226–241.

Dubins, Lester E., and David A. Freedman, 1965. A sharper form of the Borel-Cantelli lemma and the strong law. *Annals of Mathematical Statistics* **36** 800–807.

Dubins, Lester E., and Leonard J. Savage, 1960. *How to Gamble If You Must.* Privately circulated.

Dubins, Lester E., and Leonard J. Savage, 1960*a*. Optimal gambling systems. *Proceedings of the National Academy of Sciences of the United States of America* **46** 1597–1598.

Dubins, Lester E., and Leonard J. Savage, 1963. *How to Gamble If You Must: Inequalities for Stochastic Processes.* Privately circulated.

Dubins, Lester E., and Leonard J. Savage, 1965. A Tchebycheff-like inequality for stochastic processes. *Proceedings of the National Academy of Sciences of the United States of America* **51** 274–275.

Dunford, Nelson, and Jacob T. Schwartz, 1958. *Linear Operators, Part I: General Theory.* Interscience Publishers, Inc., New York.

Feldman, Dorian, 1962. Contributions to the "two-armed bandit problem". *Annals of Mathematical Statistics* **33** 847–856.

Feller, William, 1950. *An Introduction to Probability Theory and Its Applications,* Volume I. John Wiley & Sons, Inc., New York.

Gnedenko, B. V., and A. N. Kolmogorov, 1954. *Limit Distributions for Sums of Random Variables.* Translated by K. L. Chung. Addison-Wesley Publishing Company, Inc., Reading, Massachusetts.

Girshick, M. A., and L. J. Savage, 1951. Bayes and minimax estimates for quadratic loss functions. *Proceedings of the Second* [1950] *Berkeley Symposium on Mathematical Statistics and Probability.* Edited by Jerzy Neyman. University of California Press, Berkeley, California, 53–74.

Halmos, P. R., 1939. Invariants of certain stochastic transformations: The mathematical theory of gambling systems. *Duke Mathematical Journal* **5** 461–478.

Hardy, G. H., J. E. Littlewood, and G. Polya, 1934. *Inequalities.* Cambridge University Press, New York and London. (Second, similar edition 1954.)

Hausdorff, Felix, 1957. *Set Theory.* Chelsea Publishing Company, New York.

Hewitt, Edwin, 1960. Integration by parts for Stieljes integrals. *American Mathematical Monthly* **67** 419–423.

Hille, Einar, and Ralph S. Phillips, 1957. *Functional Analysis and Semi-groups,* revised edition. American Mathematical Society, Providence, Rhode Island.

Hoeffding, Wassily, 1963. Probability inequalities for sums of bounded random variables. *Journal of the American Statistical Association* **58** 13–30.

Howard, Ronald A., 1961. *Dynamic Programming and Markov Processes.* John Wiley & Sons, Inc., New York.

Hunt, G. A., 1957. Markov processes and potentials, I. *Illinois Journal of Mathematics* **1** 44–93.

Kalmár, L., 1957. Über arithmetische Funktionen von unendlich vielen Variablen, welche an jeder Stelle bloss von einer endlichen Anzahl von Variablen abhängig sind. *Colloquium Mathematicum* **5** 1–5.

Karlin, Samuel, 1955. The structure of dynamic programing models. *Naval Research Logistics Quarterly* **2** 285–294.

Kelley, John L., 1955. *General Topology.* D. Van Nostrand Company, Inc., Princeton, New Jersey.

Kemperman, J. H. B., 1961. *The Passage Problem for a Stationary Markov Chain.* The University of Chicago Press, Chicago.

König, Denes, 1936. *Theorie der endlichen und undendlichen Graphen.* Akademische Verlagsgesellschaft, Leipzig.

Kruskal, Joseph B., 1963. The number of simplices in a complex. Chapter 12, *Mathematical Optimization Techniques,* edited by Richard Bellman. Report R-366-PR, RAND Corporation, Santa Monica, California.

Kuratowski, Casimir, 1932. Les fonctions semi-continues dans l'espace des ensembles fermés. *Fundamenta Mathematicae* **18** 148–159.

Lehmer, Derrick H., 1964. The machine tools of combinatories. Chapter 1, *Combinatorial Analysis for Engineers and Scientists,* edited by Edwin F. Beckenbach. John Wiley & Sons, Inc., New York.

Lévy, Paul, 1937. *Théorie de l'Addition des Variables Aléatoires.* Gauthier-Villars, Paris. (Second edition 1954.)

Liapunov, A. A., E. A. Schegolkow, and W. J. Arsenin, 1955. *Arbeiten zur deskriptiven Mengenlehre.* Deutches Verlag der Wissenschaften, Berlin.

Loève, Michel, 1955. *Probability Theory.* D. Van Nostrand Company, Inc., Princeton, New Jersey.

Lusin, Nicolas, 1930. *Leçons sur les ensembles analytiques et leurs applications.* Gauthier-Villars, Paris.

Raiffa, Howard, and Robert Schlaifer, 1961. *Applied Statistical Decision Theory.* Graduate School of Business Administration, Harvard University, Boston, Massachusetts.

Raoult, J. P., 1964. Sur une notion de fonction universellement intégrable introduite par Dubins et Savage. *Publications de l'Institut de Statistiques de l'Université de Paris* **13** 79–108.

Raoult, J. P., 1964*a*. Résultats récents de Savage et Dubins sur les paris. *Bulletin de l'Institut des Actuaires Français* **247** 65–83.

Riley, J. A., and Theodore Hatcher, 1958. Problem E 1288. *American Mathematical Monthly* **65** 368.

Riley, J. A., and Theodore Hatcher, 1959. Problem 4858. *American Mathematical Monthly* **66** 595.

Robbins, Herbert, 1952. Some aspects of the sequential design of experiments. *Bulletin of the American Mathematical Society* **58** 527–535.

Rosenblatt, Murray, 1962. *Random Processes.* Oxford University Press, New York.

Saks, Stanislaw, 1932. *Theory of the Integral,* second revised edition. Fundusz Kultury Narodowej, Warsaw.

Savage, Leonard J., 1954. *The Foundations of Statistics.* John Wiley & Sons, Inc., New York.

SAVAGE, LEONARD J., 1957. The casino that takes a percentage and what you can do about it. Unpublished report P-1132, RAND Corporation, Santa Monica, California (July 1957, revised October 17).

SAVAGE, LEONARD J., 1962. Bayesian statistics, *Decision and Information Processes.* Edited by R. Machol and P. Gray. The Macmillan Company, New York, 161–194.

SCARNE, JOHN, 1961. *Scarne's Complete Guide to Gambling.* Simon and Schuster, Inc., New York.

SIERPINSKI, W., 1950. Les ensembles projectifs et analytiques. *Memorial des sciences mathematiques* **112** 1–80.

SPITZER, FRANK, 1956. A combinatorial lemma and its application to probability theory. *Transactions of the American Mathematical Society* **82** 323–339.

TODHUNTER, I., 1865. *A History of the Mathematical Theory of Probability from the Time of Pascal to that of Laplace.* The Macmillan Company, Cambridge and London. (Reprinted by Chelsea Publishing Company, New York, 1949.)

WALD, ABRAHAM, 1947. *Sequential Analysis.* John Wiley & Sons, Inc., New York.

WALD, ABRAHAM, 1950. *Statistical Decision Functions.* John Wiley & Sons, Inc., New York.

WALD, A., and J. WOLFOWITZ, 1948. Optimum character of the sequential probability ratio test. *Annals of Mathematical Statistics* **19** 326–339.

WEINGARTEN, HARRY, 1956. On the probability of large deviations of bounded chance variables. *Annals of Mathematical Statistics* **27** 1170–1174.

WIDDER, D. V., 1941. *The Laplace Transform.* Princeton University Press, Princeton, New Jersey.

BIBLIOGRAPHIC SUPPLEMENT (1975)

Barbosa-Dantas, C. A., 1966. *The Existence of Stationary Optimal Plans.* University of California, Berkeley. (Thesis.)

Blackwell, David, 1961. On the functional equation of dynamic programming. *Journal of Mathematical Analysis and Applications* **2** 273–276.

Blackwell, David, 1962. Discrete dynamic programming. *Annals of Mathematical Statistics* **33** 719–726.

Blackwell, David, 1964*a*. Memoryless strategies in finite-stage dynamic programming. *Annals of Mathematical Statistics* **35** 863–865.

Blackwell, David, 1965. Discounted dynamic programming. *Annals of Mathematical Statistics* **36** 226–235.

Blackwell, David, 1966. Positive dynamic programming. *Proceedings of the Fifth Berkeley Symposium on Mathematics, Statistics and Probability,* University of California, Los Angeles and Berkeley, **1** 415–418.

Blackwell, David, 1970. On stationary policies. *Journal of the Royal Statistical Society A.* **133** 33–37.

Blackwell, David, and David Freedman, 1973. On the amount of variance needed to escape from a strip. *Annals of Probability* **1** 772–787.

Blackwell, David, David A. Freedman, and M. Orkin, 1974–75. The optimal reward operator in dynamic programming. *Annals of Probability* **2** 926–941.

Chen, Robert Wen-shaing, 1974. *On Finitely Additive Almost Sure Convergence.* University of Minnesota. (Thesis.)

CHOW, Y. S., H. ROBBINS, and D. SIEGMUND, 1971. *Great Expectations: The Theory of Optimal Stopping.* Houghton Mifflin Co., Boston.

DARST, R. B., 1974. Remarks on measurable selections. *Annals of Statistics* **2** 845–847.

DEGROOT, M. H., 1970. *Optimal Statistical Decisions.* McGraw-Hill, New York.

DUBINS, LESTER E., 1966. A note on upcrossings of semimartingales. *Annals of Mathematical Statistics* **37** 728.

DUBINS, LESTER E., 1967. Subfair casino functions are superadditive. *Israeli Journal of Mathematics* **5** 182–184.

DUBINS, LESTER E., 1968*a*. A simpler proof of Smith's roulette theorem. *Annals of Mathematical Statistics* **39** 390–393.

DUBINS, LESTER E., 1968*b*. Rises of nonnegative semimartingales. *Illinois Journal of Mathematics* **12** 649–653.

DUBINS, LESTER E., 1969. On Bochner's generalization of the Radon-Nikodym theorem. *American Mathematical Monthly* **76** 520–523.

DUBINS, LESTER E., 1972*a*. On roulette when the holes are of various sizes. *Israeli Journal of Mathematics* **11** 153–158.

DUBINS, LESTER E., 1972*b*. Sharp bounds for the total variance of uniformly bounded semimartingales. *Annals of Mathematical Statistics* **43** 1559–1565.

DUBINS, LESTER E., 1972*c*. Some upcrossing inequalities for uniformly bounded martingales. *Proceedings of Symposia Matematica,* University of Rome, **IX** 169–177.

DUBINS, LESTER E., 1973. Which functions of stopping times are stopping times? *Annals of Probability* **1** 313–316.

DUBINS, LESTER E., 1974. On Lebesgue-like extensions of finitely additive measures. *Annals of Probability* **2** 456–463.

DUBINS, LESTER E., 1975. Finitely additive conditional probabilities, conglomerability and disintegrations. *Annals of Probability* **3** 89–99.

DUBINS, LESTER E., and ISAAC MEILIJSON, 1974. On stability for optimization problems. *Annals of Probability* **2** 243–255.

DUBINS, LESTER, E., and GIDEON SCHWARZ, 1965. On continuous martingales. *Proceedings of the National Academy of Sciences of the United States of America* **53** 913–916.

Dubins, Lester E., and William D. Sudderth, 1975. An example where stationary strategies are not adequate. *Annals of Probability* **3** 722–725.

Dubins, Lester E., and Henry Teicher, 1967. Optimal stopping when the future is discounted. *Annals of Mathematical Statistics* **38** 601–605.

Ferguson, Thomas S., 1965. Betting systems which minimize the probability of ruin. *Journal of the Society for Industrial and Applied Mathematics* **13** 795–818.

Freedman, David A., 1967*a*. A remark on the law of the iterated logarithm. *Annals of Mathematical Statistics* **38** 598–600.

Freedman, David A., 1967*b*. Timid play is optimal. *Annals of Mathematical Statistics* **38** 1281–1283.

Freedman, David A., 1973*a*. Some inequalities for uniformly bounded dependent variables. *Bulletin of the American Mathematical Society* **79** 40–44.

Freedman, David A., 1973*b*. Another note on the Borel-Cantelli lemma and the strong law with the Poisson approximation as a byproduct. *Annals of Probability* **1** 910–925.

Freedman, David A., 1974*a*. A remark on the strong law. *Annals of Probability* **2** 324–327.

Freedman, David A., 1974*b*. The Poisson approximation for dependent events. *Annals of Probability* **2** 256–259.

Freedman, David A., 1974–75*a*. The optimal reward operator in special classes of dynamic programming problems. *Annals of Probability* **2** 942–949.

Freedman, David A., 1974–75*b*. On tail probabilities for martingales. *Annals of Probability* **1** 100–117.

Freedman, David A., and Roger Purves, 1967. Timid play is optimal, II. *Annals of Mathematical Statistics* **38** 1284–1285.

Furukawa, Nogata, 1972. Markovian decision processes with compact action spaces. *Annals of Mathematical Statistics* **43** 1612–1622.

Heath, David, and William D. Sudderth, 1974. Continuous-time gambling problems. *Advances in Applied Probability* **6** 1–14.

Heath, David, William Pruitt, and William D. Sudderth, 1972. Subfair Red-and-Black with a limit. *Proceedings of the American Mathematical Society* **35** 555–560.

HEWITT, EDWIN, and LEONARD J. SAVAGE, 1955. Symmetric measures on Cartesian products. *Transactions of the American Mathematical Society* **80** 470–501.

HINDERER, KARL F., 1970. Foundations of non-stationary dynamic programming with discrete time parameter. *Lecture Notes In Operations Research and Mathematical Systems* **33**. Springer Verlag, Berlin.

JACOBS, CONRAD, 1969. Rot und Schwarz. *Selecta Mathematica* **1**. *Heidelberger Taschenbücher* **49**. Springer-Verlag, Berlin, New York.

MAITRA, A., 1965. Dynamic programming for countable state systems. *Sankhya* **27A** 241–248.

MAITRA, A., 1968. Discounted dynamic programming on compact metric spaces. *Sankhya* **30A** 211–219.

MACQUEEN, JAMES B., 1961. A problem in survival. *Annals of Mathematical Statistics* **32** 605–610.

MACQUEEN, JAMES B., 1962. Further problems in individual and group survival. Working Paper No. **20**. *Western Management Science Institute,* University of California, Los Angeles.

MACQUEEN, JAMES B., 1964. A problem in making resources last. *Management Science* **11** 341–347.

MACQUEEN, JAMES B., 1965. Some methods for classification and analysis of multivariate observations. *Proceeding of the Fifth Berkeley Symposium on Mathematical Statistics and Probability,* University of California Press, 281–297.

MACQUEEN, JAMES B., 1973. A linear extension of the martingale convergence theorem. *Annals of Probability* **1** 263–271.

MEYER, PAUL ANDRÉ, and M. TRAKI, 1971–72. Reduites et Jeux de Hasard. Seminaire de Probabilites VII, Universite de Strasbourg, *Lecture Notes In Mathematics* **321** 155–171. Springer-Verlag, Berlin.

ORNSTEIN, DONALD, 1969. On the existence of stationary optimal strategies. *Proceedings of the American Mathematical Society* **20** 563–569.

PRABHAKAR, N. D., 1972. *A Study of Dynamic Programming and Gambling Systems.* Indian Statistical Institute, Calcutta. (Thesis.)

PURVES, ROGER, and WILLIAM D. SUDDERTH, 1976. Some finitely

additive probability. Submitted to *Annals of Probability. Statistical Technical Report 220,* University of Minnesota.

SIRJAEV, A. N., 1970. Some new results in the theory of controlled random processes. *Selected Translations in Mathematical Statistics and Probability* **8** 49–130.

SMITH, GERALD, 1967. Optimal strategy at roulette. *Zeitschrift fur Wahrscheinlichkeitstheorie und Verwandte Gebiete* **8** 91–100.

SNELL, J. L. 1952. Applications of martingale system theorems. *Transactions of the American Mathematical Society* **73** 293–312.

STRAUCH, RALPH E., 1966. Negative dynamic programming. *Annals of Mathematical Statistics* **37** 871–890.

STRAUCH, RALPH E., 1967*a*. Measurable gambling houses. *Transactions of the American Mathematical Society* **126** 64–72.

STRAUCH, RALPH E., 1967*b*. Correction. *Transactions of the American Mathematical Society* **130** 184.

SUDDERTH, WILLIAM D., 1969*a*. A note on thrifty strategies and martingales in a finitely additive setting. *Annals of Mathematical Statistics* **40** 2211–2214.

SUDDERTH, WILLIAM D., 1969*b*. On measurable, nonleavable gambling houses with a goal. *Annals of Mathematical Statistics* **40** 66–71.

SUDDERTH, WILLIAM D., 1969*c*. On the existence of good stationary strategies. *Transactions of the American Mathematical Society* **135** 399–414.

SUDDERTH, WILLIAM D., 1971*a*. A "Fatou equation" for randomly stopped variables. *Annals of Mathematical Statistics* **42** 2143–2146.

SUDDERTH, WILLIAM D., 1971*b*. A gambling theorem and optimal stopping theory. *Annals of Mathematical Statistics* **42** 1697–1705.

SUDDERTH, WILLIAM D., 1971*c*. On measurable gambling problems. *Annals of Mathematical Statistics* **42** 260–269.

SUDDERTH, WILLIAM D., 1972. On the Dubins and Savage characterization of optimal strategies. *Annals of Mathematical Statistics* **43** 498–507.

TAYLOR, HOWARD M., 1972. Bounds for stopped partial sums. *Annals of Mathematical Statistics* **43** 733–747.

Turnbull, Bruce W., 1973. Inequalities for branching processes. *Annals of Probability* **1** 457–474.

Wilkins, J. Ernest, 1972. The bold strategy in presence of house limit. *Proceedings of the American Mathematical Society* **32** 567–570.

INDEX OF PERSONS

Works cited are listed in the Bibliography under first or only author.

SUBJECT INDEX

There are also an Index of Persons (pp. 239–240), an Index of Symbols (p. 247), and an Index of Open Problems (p. 249).

Defining contexts are here indexed with italicized page numbers.

INDEX OF SYMBOLS

Defining contexts of some frequently used technical symbols are indexed here in order of occurrence. Related symbols introduced on the same page are grouped.

INDEX OF OPEN PROBLEMS

References to problems left open are here indexed in order of occurrence.